BOGATIN'S PRACTICAL GUIDE to PROTOTYPE BREADBOARD and PCB DESIGN

ERIC BOGATIN

E-ISBN: 978-1-63081-848-7
Cover design by Charlene Stevens

© 2021
Artech House
685 Canton Street
Norwood, MA 02062

ARTECH HOUSE
BOSTON | LONDON
artechhouse.com

Table of Contents

Chapter 1	A Getting-Started Guide	13
1.1	Who This Book Is For	14
1.2	Getting Stuff Done	16
1.3	Cost-Performance Trade-offs	18
1.4	Errors, Best Practices, and Habits	22
1.5	Learn to Design-in Success	25
1.6	A Getting-Started Guide for Signal Integrity	26
1.7	The Seven-Step Process	29
1.8	Risk Management and Mitigation	30
1.9	Two Risk Management Design Strategies	32
1.10	Master of Murphy's Law	33
1.11	Proof of Concept	35
1.12	Practice Questions	38
Chapter 2	PCB Technology	39
2.1	PCB, PWB, or PCA?	39
2.2	Physical Design of a PCB	40
2.3	Vias Technologies	41
2.4	Thermal and Thermal Relief Vias	45
2.5	Other Layers	48
2.6	The Soldermask Layer	49
2.7	Surface Finishes	51
2.8	The Silk Screen	53
2.9	What the Fab Vendor Needs	54
2.10	Practice Questions	55
Chapter 3	Signal Integrity and Interconnects	57
3.1	Transparent Interconnects	58

3.2	When Interconnects are NOT Transparent	60
3.3	Where Signal Integrity Lives	65
3.4	Six Categories of Electrical Noise	71
3.5	Families of SI/PI/EMI Problems	74
3.6	In Principle and In Practice	76
3.7	Net Classes and Interconnect Problems	79
3.8	Design for Performance	82
3.9	Design for X	84
3.10	Practice Questions	85
Chapter 4	Electrical Properties of Interconnects	87
4.1	Ideal vs Real Circuit Elements	87
4.2	Equivalent Electrical Circuit Models	94
4.3	Parasitic Extraction of R, L, and C Elements	98
4.4	Describing Cross Talk	102
4.5	Estimating Mutual Inductance	105
4.6	Training Your Engineer's Mind's Eye	108
4.7	Electrically Long Interconnects	109
4.8	Electrically Short and Electrically Long	111
4.9	Practice Questions	119
Chapter 5	Trace Width Considerations: Max Current	121
5.1	Best design practices	122
5.2	Minimum Fabrication Trace Width	122
5.3	Copper Thickness as Ounces of Copper	123
5.4	Maximum Current Handling of a Trace	126
5.5	Maximum Current Through a Via	132
5.6	Thermal Runaway with Constant Current	133
5.7	Practice Questions	140
Chapter 6	Trace Width Considerations: Series Resistance	143

6.1	Resistance of Any Uniform Conductor	143
6.2	Sheet Resistance of a Copper Layer	148
6.3	Measuring Very Low Resistances	152
6.4	Voltage Drop Across Traces	156
6.5	The Thevenin Model of a Voltage Source	157
6.6	How Much Trace Resistance Is too Much?	165
6.7	The Resistance of a Via	167
6.8	Resistance of a Thermal Relief Via	172
6.9	Practice Questions	174
Chapter 7	The Seven Steps in Creating a PCB	177
7.1	Step 1: Plan of Record	178
7.2	Step 2: Create the BOM	180
7.3	Step 3: Complete the Schematic	182
7.4	Step 4: Complete the Layout, Order the Parts	183
7.5	Steps 5 and 6: Assembly and Bring-Up	185
7.6	Step 7: Documentation	187
7.7	Practice Questions	189
Chapter 8	Step 1, POR: Risk Mitigation	191
8.1	Visualize the Entire Project Before You Begin	191
8.2	Avoid Feature Creep	193
8.3	Estimate Everything You Can	193
8.4	Preliminary BOM: Critical Components	195
8.5	Risk Assessment	195
8.6	Risk Mitigation: Tented Vias	198
8.7	Risk Mitigation: Qualified Parts	202
8.8	Practice Questions	204
Chapter 9	Risk Reduction: Datasheets, Reverse Engineering, and Component Selection	207

9.1	Take Responsibility for Your Design	207
9.2	Reducing the Risk of a Design Problem	208
9.3	Understand Your Circuit	209
9.4	Read Datasheets Critically	212
9.5	Build Simple Evaluation Prototypes	212
9.6	Reverse Engineer Components	213
9.7	Reuse Parts	216
9.8	Practice Questions	218
Chapter 10	Risk Reduction: Virtual and Real Prototypes	221
10.1	Getting Started with Circuit Simulation	221
10.2	Practice Safe Simulation	225
10.3	Simulating a 555 Circuit	227
10.4	Purchase an Evaluation Board	231
10.5	Real Prototypes with Modules	231
10.6	Practice Questions	233
Chapter 11	Risk Reduction: Prototyping with a Solderless Breadboard	235
11.1	Build a Real Prototype	236
11.2	Solderless Breadboards for POC	238
11.3	Features of a Solderless Breadboard	239
11.4	Bandwidth Limitations	244
11.5	A Simple Breakout Board	251
11.6	The Mini Solderless Breadboard	254
11.7	Best Wiring Habits	255
11.8	Habit #1: Consistent Column Assignments	256
11.9	Habit #2: Color Code the Wires	259
11.10	Habit #3: Keep Signal Traces Short	260
11.11	Habit #4: Avoid a Shared Return Path	263
11.12	Habit #5: Route Signal-Return Pairs	263

11.13	Habit #6: Keep Component Leads Short	266
11.14	Practice Questions	267

Chapter 12   Switching Noise and Return Path Routing 269

12.1	The Origin of Switching Noise	271
12.2	Signal-Return Path Loops	274
12.3	Where Does Return Current Flow?	279
12.4	A Plane as a Return Path	283
12.5	Ground	287
12.6	Avoid Gaps in the Return Plane	290
12.7	Summary of the Best design practices	292
12.8	Practice Questions	293

Chapter 13   Power Delivery 295

13.1	Origin of Power Rail Switching Noise	295
13.2	Calculating Loop Inductance	298
13.3	Measuring PDN Switching Noise	300
13.4	The Role of Decoupling Capacitors	302
13.5	Where Do Decoupling Capacitors Go?	305
13.6	The Power Delivery Path	309
13.7	Inrush Current	311
13.8	Summary of the Eight Habits for Using a SSB	312
13.9	Practice Questions	313

Chapter 14   Design for Performance: The PDN on a PCB 315

14.1	VRM specifications	316
14.2	Voltage Regulator Module	317
14.3	Self- and Mutual-Aggression Noise	319
14.4	Power and Ground Loop Inductance	320
14.5	Decoupling Capacitors	323
14.6	A Decoupling Capacitor Myth; Part 1	326

14.7	A Decoupling Capacitor Myth; Part 2	330
14.8	Routing for Power Distribution	335
14.9	Ferrite Beads	336
14.10	Summary of the Best design practices	343
14.11	Practice Questions	344
Chapter 15	Risk Reduction: Design for Bring-Up	347
15.1	Test is Too General a Term	347
15.2	What Does It Mean to "Work"?	351
15.3	Design for Bring-Up	353
15.4	Add Design for Bring-Up Features	355
15.5	Jumper Switches	358
15.6	LED indicators	359
15.7	Test Points	361
15.8	The Power Rail as a Diagnostic	366
15.9	Practice Questions	371
Chapter 16	Risk Reduction: Design Reviews	373
16.1	The Preliminary Design Review	373
16.2	The Critical Design Review	374
16.3	DRC for DFM in the CDR	378
16.4	DRC for Signal Integrity	378
16.5	Layout Review	379
16.6	Practice Questions	381
Chapter 17	Step 2: Surface-Mount or Through-Hole Parts	383
17.1	Through-Hole and Surface-Mount	383
17.2	Types of SMT Parts	385
17.3	Integrated Circuit Components	388
17.4	Practice Questions	393
Chapter 18	Finding the One Part in a Million	395

18.1	An Important Selection Process	395
18.2	Trade-offs in Selecting Parts	397
18.3	The Search Order to Select a Part	398
18.4	Selecting Resistors	402
18.5	Selecting Capacitors	405
18.6	The BOM	408
18.7	Summary of the Best design practices	409
18.8	Selecting Parts for Automated Assembly	411
18.9	Practice Questions	412
Chapter 19	Step 3: Schematic Capture and Final BOM	413
19.1	Picking a Project Name	414
19.2	Schematic Capture	416
19.3	Take Ownership of Reference Designs	417
19.4	Add Options to Your Schematic	418
19.5	Best design practices for Schematic Entry	418
19.6	Design Review and ERC	423
19.7	Practice Questions	424
Chapter 20	Step 4: Layout — Setting Up the Board	425
20.1	Layout	425
20.2	Board Dimensions	426
20.3	The Layers in a Board Stack	428
20.4	Negative and Positive Layers	428
20.5	Examples of Some Fab Shop DFM Features	430
20.6	Setting Up Design Constraints	432
20.7	Thermal Reliefs in Pads and Vias	433
20.8	Set Up Board Size and Keepout Layer	436
20.9	Practice Questions	437
Chapter 21	Floor Planning and Routing Priority	439

21.1	Part Placement	439
21.2	The Order of Placement and Routing	442
21.3	First Priority: Ground Plane on the Bottom Layer	443
21.4	Second Priority: Decoupling Capacitors	445
21.5	Third Priority: Ground Connections	446
21.6	Fourth Priority: Digital Signals, Congested Signals	447
21.7	Fifth Priority: Power Paths	449
21.8	The Silk Screen	451
21.9	Check the Soldermask	453
21.10	Soldermask Color	454
21.11	Layout — Critical Design Review	455
21.12	Practice Questions	458

Chapter 22	Six Common Misconceptions about Routing	459
22.1	Myth #1: Avoid 90 Deg Corners	459
22.2	Myth #2: Add Copper Pour on Signal Layers	461
22.3	Myth #3: Use Different Value Decoupling Capacitors	465
22.4	Myth #4: Split Ground Planes	465
22.5	Myth #5: Use Power Planes	466
22.6	Myth #6: Use 50 Ohm Impedance Traces	467
22.7	Practice Questions	470

Chapter 23	Four-Layer Boards	473
23.1	Two-Layer Stack-Ups	473
23.2	A 4-Layer Board	474
23.3	Four-Layer Stack-Up Options	476
23.4	Stack-Up Options with Two Planes	478
23.5	The Recommended 4-Layer Stack-Up	480
23.6	When Signals Change Return Planes	481
23.7	Practice Questions	483

Chapter 24	Release the Board to the Fab Shop	485
24.1	Gerber Files	485
24.2	Cost Adders	486
24.3	Board Release Checklist	488
24.4	Practice Questions	489
Chapter 25	Step 6: Bring-Up	491
25.1	Does Your Widget Work?	491
25.2	Prototype or Production Testing	492
25.3	Design for Bring-up	494
25.4	Find the Root Cause	495
25.5	Problems to Expect	497
25.6	Troubleshoot Like a Detective	500
25.7	Trick #1: Recreate the Problem	502
25.8	Trick #2: Seen This Problem Before?	503
25.9	Trick #3: Round Up the Usual Suspects	504
25.10	Trick #4: Three Possible Explanations	505
25.11	A Methodology	506
25.12	Forensic Analysis	507
25.13	Coding Issues	512
25.14	Practice Questions	513
Chapter 26	Step 7: Documentation	515
Chapter 27	Concluding Comments	517
Chapter 28	About Eric Bogatin	519

Dedication

This book is dedicated to my wife, Susan. Her constant encouragement and support made this book possible.

Chapter 1
A Getting-Started Guide

Printed circuit boards range in complexity from a simple 1-layer board with a few components and nets, to a 32-layer or more board with thousands of components and tens of thousands of nets. This range is illustrated in **Figure 1.1.**

Figure 1.1 An example of the range of circuit board complexity. Left: a simple 1-layer board, courtesy of Altium and right: a complex server board, courtesy of Intel.

In designing, fabricating, and bringing to life these boards, there are some basic techniques we will use for both the simplest and the most complex boards and there are some techniques we will need only for the most complex boards.

This book is really a getting started guide in the basic techniques we will use over and over again for all circuit boards. We limit the scope of this book to simpler, 2-4-layer designs that can generally be purchased for the lowest prices on internet PCB fab sites and assembled remotely using captive assembly lines with limited, captive parts stockrooms.

When a completely assembled, fully custom board can be purchased for less than $10 a board in quantities of two, these skills will enable a hobbyist, a maker, a student, or a young engineer to accelerate up the learning curve as an experienced hardware engineer.

1.1 Who This Book Is For

This book is for the casual user, not the power user. We focus primarily on the outcome of the design process, and only a little on the specific mouse clicks or menu items to implement the outcome, which will vary from electronic design automation (EDA) tool to EDA tool.

For example, when placing a via to route a signal from one layer to another, there are multiple methods. Regardless of the tool, unless you have a strong compelling reason otherwise, you should always use a via with the smallest drill diameter your fab vendor can implement with no cost adder.

In a 4-layer board, a return via should be added adjacent to the signal via if the return plane changes.

These are design principles you will learn to implement as a good habit to reduce noise. *What* you should implement in your designs as good habits is the focus of this book; *how* you implement the features using your favorite EDA tool is still very important as a designer but is not the focus of this book.

The methods we introduce in this book can be implemented with the range of EDA PCB design tools, from the free online internet tools to the high-end professional tools.

If you are looking for a tool to use, there are generally two classes: simple and free, and professional level and not free.

Simple, low-end, free tools:

- ✓ Circuit Maker, from Altium: https://circuitmaker.com/
- ✓ Fritzing: https://fritzing.org/
- ✓ Easyeda: https://easyeda.com/editor

- ✓ DesignSpark: https://www.rs-online.com/designspark/home
- ✓ PCB123, from Sunstone: https://www.sunstone.com/pcb-resources/cad-tools/pcb123

Professional-level tools for engineers:

- ✓ KiCAD: https://kicad.org/ (free)
- ✓ Eagle from Autodesk: https://www.autodesk.com/products/eagle/overview?plc=F360&term=1-YEAR&support=ADVANCED&quantity=1
- ✓ Altium Designer: https://www.altium.com/
- ✓ Siemens PADS Pro: https://www.pads.com/
- ✓ Cadence Allegro: https://www.cadence.com/en_US/home/tools/pcb-design-and-analysis/pcb-layout/allegro-pcb-designer.html

Whichever tool you use, learn the basic steps to capture a schematic and design a board layout. The more you use the tool, the farther up the learning curve you will want to advance to become more efficient with your chosen tool.

There are some limitations to the type of designs we cover in this book. We assume all the routing can be implemented in an up to 4-layer board using at most two signal layers. This means the challenges of ball grid array (BGA) parts are not covered in this book. These components require a whole new set of design and layout principles because of the higher interconnect densities required.

If you deal with BGA parts, you will want to check out the book by Charles Pfeil, *BGA Breakouts And Routing: Effective Design Methods For Very Large BGAs*, available here, for example: https://www.amazon.com/BGA-Breakouts-Routing-Effective-Methods/dp/1452867585

While the examples in this book are implemented as 2- and 4-layer boards, the driving forces for the guidelines and habits recommended will apply across the spectrum of boards.

The most basic electrical considerations and analysis for current handling, return path control, power rail decoupling, and ground bounce reduction are introduced and will transcend all designs.

The principles underlying the design guidelines to reduce electrical noise are extensively included so that every designer will be empowered to make their own decisions when trading off design features with acceptable noise, cost, schedule, and risk.

The habits introduced in this book will apply from DC to very high frequency, but they are not enough.

We limit the electrical considerations to those interconnects that are electrically short: transmission line behavior does not apply, interconnects do not require termination strategies, and the routing topologies do not affect signal quality. These are very important considerations when interconnects are electrically long and this topic is covered in detail in another book in this series, Bogatin's Practical Guide to Transmission Line Design and Characterization for Signal Integrity Applications, published by Artech in May 2020.

A functioning circuit board is more than just a bare printed circuit board. It also includes the component selection, circuit design, assembly and bring-up, test, and characterization. These topics, especially when the lowest cost is an important driving force, are covered extensively in this book.

1.2 Getting Stuff Done

This book focuses on GSD (getting s#!t done). It is a practical, how-to, soup-to-nuts book on how to take a back of the napkin (BoN) sketch and turn it into a working widget prototype. An example is shown in **Figure 1.2**.

1.2 Getting Stuff Done 17

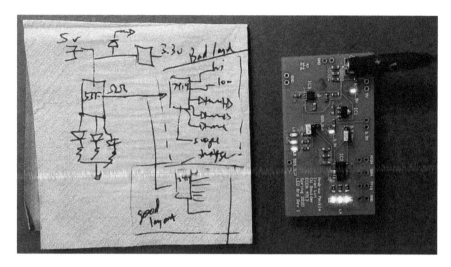

Figure 1.2 An example of a back of the napkin (BoN) sketch for a concept and the working widget prototype from this concept, built by one of my students.

This book introduces a seven-step process for prototype development. Once the prototype has been verified, there is a different development track to turn this prototype into a production-ready product and then go through qualification testing, certification testing, ramping up volume production, and cost reductions.

The production ramp path is outside the scope of this book.

The purpose of this book is to increase your success at turning a BoN sketch into a working widget prototype when a printed circuit board is required as the backbone platform to interconnect all your components.

More importantly, it is about developing good design and measurement habits, which will increase your chance of success in all of your projects and position your circuit board projects for production. You will use these habits over and over again, for all of your designs. Hopefully, you will teach others these good habits so that your team will have an increased chance of success as well.

1.3 Cost-Performance Trade-offs

The most important goal in product design, which influences all decisions you make, is to achieve a target performance (provide value to the final customer) at an acceptable price to the final user and at an acceptable cost to produce.

When profit is important, the goal is maximum value to the customer and profit to the company. When profit is not important, such as building a prototype for internal use, the goal is maximum value at an acceptable cost.

In the development phase, the cost is not just the final cost to manufacture the product. It also includes the development costs in terms of cash, schedule, and risk. This fundamental trade-off is illustrated in **Figure 1.3**.

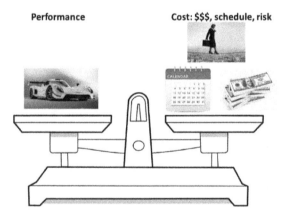

Figure 1.3 The fundamental trade-offs between performance and the three different costs in the development phase.

Sometimes it is worth it to pay more for a component if it reduces the total cost of ownership (TCOO). For example, using an integrated switch mode power supply (SMPS) may be more expensive than the discrete solution with multiple resistors, capacitors, and inductors, but the integrated solution may reduce assembly cost and risk. This is part of the trade-off decision.

1.3 Cost-Performance Trade-offs

Sometimes, it is worth adding a higher cost feature if it gives your product higher value. Adding a Bluetooth interface may enable your product to be controlled by a smart phone. If the extra cost adds enough value to cover this cost and some profit, it has a favorable return on investment (ROI). This is part of the trade-off decision.

Rarely is it possible to add value to the product and not increase one of the cost metrics. When you can increase value with little cost adder, it's a no-brainer. Make the decision, as long as there is no strong compelling reason otherwise.

Using a soldermask color on your circuit board that adds brand recognition to your product, like the purple of Adafruit or the red of Sparkfun, is a no-brainer.

Increasing the temperature resolution on a display from 0.1 deg to 0.01 deg at the cost of a $1.05, 16 bit ADS1115 ADC may not be worth the value returned for an ambient room temperature sensor, but may be well worth it for a health monitor temperature sensor.

Rarely is it possible to reduce one of the cost metrics without increasing one of the other metrics. Using six layers for your circuit board may reduce the risk of a noise problem but will increase the dollar cost budget.

In the product definition stage, this trade-off is sometimes referred to as the triple constraint between:

- ✓ Cost
- ✓ Time
- ✓ Scope (features or value)

Once the product is defined, in the product development phase, this is referred to as the constraint triangle:

- ✓ Lowest cost
- ✓ Highest quality
- ✓ Done quickly

You can have any two, but not all three at the same time.

The product development process is really about balancing these sorts of trade-offs.

There is an important, explicit element missing from these analyses that applies specifically to new product development and prototypes in particular: the *risk* of not being successful.

The element of risk is buried in the quality statement but should be explicitly brought out and addressed. Whenever you are doing something new that has not been done before, in exactly the way you are doing it, there is always the element of risk:

- ✓ Will the components you select work the way you expect?
- ✓ Will the circuit schematic be correct?
- ✓ Will you be able to create the code needed?
- ✓ Will the circuit board footprints match the pads of your components?
- ✓ Will your board come back from the vendor on the schedule you expect?
- ✓ Will the noise created by the circuit board be low enough to not interfere with the operation of the circuit?

Risk reduction rarely comes for free. Paying more dollars for features to reduce noise will reduce risk or reduce the time if it means no second board spin. Sometimes, it is worth paying more for reduced risk and reduced time. This is called *buying insurance*.

If you reduce the dollar cost, you may increase the risk of the product not working initially (the *quality*) or over the product lifetime (the *reliability*). If you buy a lower-cost row of header pins from a new vendor and it breaks too easily, it may not be worth the cost reduction. This is why it is important to qualify a part and a vendor before you commit to it. This is part of risk reduction.

If you reduce the schedule, you may have to increase the risk and increase the dollar cost. If you do not want to take the time for a qualification test of a new part, you can pay more for an unqualified part from a qualified vendor, or use an unqualified part in stock rather than wait for your qualified part.

If you can make a design decision that reduces the risk without impacting performance or value, at the same or lower cost or schedule, it's a no-brainer. We call these decisions, *best design practices*.

> A best design practice is a decision you make that reduces risk, or reduces cost or development time, with a favorable return on investment. Always follow a BDP unless you have a strong compelling reason otherwise.

A little upfront planning to avoid common risks can have a large impact on success, and a favorable ROI. This is a best design practice.

Using a ground plane on the bottom layer of a circuit board will reduce some sources of noise. If you can fit all or most of your interconnects on the top layer, using the bottom layer for a continuous ground has a favorable ROI. This is a best design practice.

Using silk screen labels to clearly identify the test points on your board for the 5 V rail, the clock output, and the three critical signals, does not add cost but will reduce the time for bring-up and the risk of measuring the wrong signal. It is a no-brainer and has a favorable ROI. This is a best design practice.

When it costs more if you follow a best design practice, it's time to bring out your engineer's hat and sharpened pencil and paper and do a cost-performance trade-off analysis. This usually involves estimating the impact on performance, risk, cost, and schedule using rules of thumb, approximations, or numerical simulations.

Using a more expensive capacitor rated for 200 V will be more reliable than a similar-value, lower-cost capacitor, but rated at only 25 V. But if the application is for a 5 V rail, the lower reliability risk offers no additional value so this decision may not be a favorable ROI.

In some companies, historical events affect what is perceived as acceptable risk. If a significant problem resulted in a previous product's delayed introduction to the market, the fix may be required for all future product, whether it is needed or not. This is an emotional response to an engineering trade-off.

If a ceramic resonator was used as a clock reference and it resulted in a batch of microcontroller boards that did not boot because the resonator did not oscillate, the corporate culture may ban ceramic resonators in favor of only quartz crystal resonators. This may be in spite of the fact that the resonator circuit just needed a shunt resistor to self-start. Once burned by a costly mistake, management has a tendency to overcorrect in future designs, independent of what is a good engineering decision.

1.4 Errors, Best Practices, and Habits

There are many decisions you will make when designing a prototype. Should I use a 2-layer board or a 4-layer board? Will this low drop out (LDO) 3.3 V regulator work in my application? Should I use a 0.1 uF and 10 uF capacitor on my 5 V rail or a 22 uF capacitor? Should I use the letters "TP1" for the silk screen label for a test point, or "5V rail" as the label?

There are many more ways of designing a product that fails than designing one that works. Navigating the decision tree and balancing all the trade-offs when designing and building your prototype is what engineering is all about.

There are generally two types of errors you can make that result in some form of reduced value or increased cost:

- ✓ Hard errors: the product does not work
- ✓ Soft errors: the product works, but the cost or noise (risk) may be higher than it could be

Sometimes, there are multiple ways of implementing a feature that will have a neutral impact on value or cost. The decision choice is a *personal preference*. These are not errors, but more about style. You choose which path to take.

Some decisions you make will result in a *hard error*. Your prototype will not work due to this problem. Using a 6 mil wide trace to route 10 A of current to a motor will result in a catastrophic failure when the trace melts or catches fire. This would be a hard error. Forgetting to connect a ground connection to an input power jack will result in your board not being powered when plugged in. This is a hard error.

Routing the TX from the controller to the TX pin of the UART driver will cause the communications link to fail. This is a hard error.

The other technical term for a hard error is, "you're screwed."

Some decisions you make will be neutral. Your product will probably work *in spite* of your decisions. If you had the chance to do it differently next time, you would change your decision. This is a *soft error*.

Routing a long trace in the bottom ground layer may increase the ground bounce noise in your design, but it may still work, in this board, just fine. This is a higher risk and should be avoided next time.

The decoupling capacitor is placed an inch away from the power pin it is decoupling. It will probably work in this design, but there may be some combinations of microcode and functions where a bit error occurs. This is very hard to diagnose. It is a soft error and should be avoided in the future.

Not including a test point for the 5 V power rail will make the debugging more difficult, but the product will probably still work. This will increase the risk and make it harder to debug your board. It is a soft error.

Labeling a test point as TP31 means you have to spend the time looking up what signal is at this test point. Next time, you would label this as 555 output, for example. This is a soft error.

Not adding your name or board ID to the silk screen is inconvenient, especially if you have ten different board designs you are exploring on your lab bench. The chance of getting them confused is high. Next time you would add a unique board identifier in silk screen. This is a soft error. Of course, if it means not getting your board back from the general pool of boards, it could be a hard error.

Eliminating the hard errors and reducing the soft errors that may creep into your design is about reducing risk. An important process for reducing the risk of an error is by following best design practices.

There are so many little mistakes that are easy miss in a design, any of which could be either a hard or soft error. An important best design practice to minimize the chance of a mistake not being caught before the product is released to fab is to keep a checklist of possible problems. Many checklists are provided in this book. Start with these and add your own mistakes to the list. We will leverage the phrase over and over again.

> *An expert is someone who has made all the mistakes possible.*

However, it is not necessary for you personally to make all the mistakes possible. Learn from the mistakes others have made.

Generally, we will try to use best design practices that can be leveraged across many designs and many applications. Learning them once, you will use them over and over again. If they are free and do not come at additional cost, they should become *habits*.

For example, an important best design practice is to use a continuous ground plane on the layer adjacent to signal layers. This best design practice will reduce the risk of ground bounce noise, a type of cross talk and a type of switching noise, which is often a source of unexpected product failure.

Another best design practice is to use net names on all wires using a labeling scheme that groups similar nets in order when listed alphabetically. This simple best design practice reduces the risk of forgetting a connection or making the wrong connections when doing the board layout.

Another best design practice is to use the minimum line width your fab shop can fabricate in order to increase your routing interconnect density without adding costs. Keep in mind, a 6 mil wide trace can handle as much as 1 A of current.

If the best design practice has a favorable ROI and is free, it is something you should always do, unless you have a strong compelling reason otherwise. We call these *habits*.

A habit is something you do without thinking. It comes automatically. And it will increase your chance of success with little cost impact.

When the best design practice becomes a habit, you will be a more effective engineer. Teach these habits to the rest of your team and you will collectively become a more effective team. But recognize their origin so you can evaluate when you may have a strong compelling reason to not follow the habit.

1.5 Learn to Design-in Success

The traditional approach of prototype development is to "build it and test it." When it doesn't work, try to figure out why, and usually before you can find all the problems, the pressure is on to redesign it, build it again, test it, and see how far you can get.

Sometimes, this approach is called "testing-in success." This approach is way too inefficient. It takes too long, costs too much, and does not guarantee success.

The alternative is to plan to get it right the first time. And when you don't get it right the first time, have a strategy to get it right the second time.

> *Spending an extra day in the planning phase may save you weeks to complete a new board design-spin, even assuming you were able to find the actual design features to change to get it right in the next release.*

It is in the planning phase that you should anticipate as many of the potential problems that can arise and put in place a plan to avoid each one. This is where your checklist of common mistakes made by others and which you have made is a valuable mechanism to not repeat them.

> *Use every mistake you make to ratchet up the learning curve so you never repeat it.*

Before this book, your product development strategy should have been: design and execute the product as well as you can, but anticipate there will be problems and unknowns, so plan to *fail early and fail often* and learn from your mistakes.

Following the guidelines in this book, your product development strategy should be *fail less, succeed more.*

> *This book introduces an efficient process to integrate best design practices to get it right the first time and best measurement practices to get it right the second time.*

1.6 A Getting-Started Guide for Signal Integrity

This book is a getting-started manual for any engineering student or professional engineer to develop good design and measurement habits when designing and building electronic prototypes.

The schematic describes all of the components used in a design and how they are connected. On each component, like a resistor, capacitor, LED, connector, op-amp, transistor, or other IC are connection points or *terminals* or *pins* or *nodes*. These are the features of each component electrically connected to other components.

All of this information about the components and their terminals is contained in a database for the bill of materials (BOM).

1.6 A Getting-Started Guide for Signal Integrity 27

All terminals that are connected together are connected by one specific *net*. The list of all the nets and what terminals are connected to each net is called the net list. This is the second important database that really defines a design.

In a sense, the schematic is a graphical representation of the BOM and the net list databases. In most electronic design automation (EDA) tools, we manipulate the BOM and net list databases through a graphical user interface, the schematic capture tool.

The connections between the terminals of each component in a schematic are meant to only represent connectivity. They have no electrical properties other than an assumed 0 ohm connection. They say nothing about electrical performance.

It is when the schematic is translated into the physical interconnects of traces of copper that the electrical properties of the copper wires will play a role. All of the real, physical conductor (and insulating) structures that actually perform the electrical connections are considered the *interconnects*. All the cables, connectors, and any other conductors included in the connection paths from terminal to terminal, are considered the interconnects. Even the dielectrics that provide isolation between conductors are part of the interconnects.

When interconnects are electrically transparent, the interconnects' only purpose is to provide *connectivity* between the terminals of all the components in the product. This is the case in both prototype solderless breadboard circuits and prototype printed circuit boards.

Once connectivity is established, the interconnects can only increase the electrical noise. Being transparent means the noise generated is acceptable *for that design*. But as rise times decrease and signals become more sensitive to noise, even in many common, low-cost designs, the interconnects may no longer be transparent. Their impact on signal quality must be taken into account in the design phase or your product may not work.

To evaluate when interconnects are no longer transparent, the very first step is to translate the physical properties of the interconnects, and their geometry and material properties, into their equivalent electrical circuit elements. Since the impact from these circuit elements is only to "eat design margin" and "degrade performance," the equivalent electrical circuit elements of the interconnects are sometimes referred to as the *parasitics* of the interconnects.

Based on the equivalent circuit models of the interconnects, we can use rules of thumb, simple estimates, and numerical simulations to anticipate or predict the impact on noise from the interconnect design. Knowing the types of problems that can arise, we can establish a set of best design practices that will decrease the noise generated by the parasitics.

The process of translating the physical interconnects into their parasites, analyzing the impact from these parasitics, establishing design practices to reduce their impact, and the best measurement practices to measure these effects is all encompassed by the term *signal integrity*.

> *This book establishes the essential principles to get started with signal integrity and the best practices to mitigate these effects to end up with an acceptable design.*

This book covers the five important areas:

- ✓ Essential principles at the foundations of signal integrity
- ✓ How to translate physical interconnects into equivalent electrical circuit elements (parasitic extraction)
- ✓ How to estimate the impact on noise from the parasitics
- ✓ Best design practices for acceptable signal integrity
- ✓ Best measurement practices for signal integrity

If you have been designing interconnects assuming they are transparent, this book will recalibrate your thinking to view interconnects as electrical circuit elements.

If you have been following design guidelines handed down from generation to generation, you will see some of the origins of these recommendations and make your own judgment as to when they apply and when they don't.

With the principles introduced in this book, you will be empowered to be your own expert.

However, this book is just a getting-started guide. There are many more advanced signal integrity principles and best design and measurement principles to consider for ever-higher-speed products. The principles described in this book are at the foundation of everything else you will want to do in every future higher-performance product.

Even if, in your current design, the interconnects are transparent, the habits you learn for good signal integrity design will position you for your next product where the interconnects may not be transparent. When the best design and measurement practices become habits, it is more likely you will implement them in all of your designs and increase the chance of success in all of your future product designs.

1.7 The Seven-Step Process

Every new project starts with the product concept and the rough functional or block diagram or schematic that could be sketched on a napkin. In almost every case, we turn this concept into a working prototype circuit board by following these 7 steps:

1. Complete the plan of record (POR).
2. Complete the preliminary bill of materials (BOM).
3. Complete the final schematic capture and final BOM.
4. Complete the board layout and order all the parts.
5. Complete the assembly.
6. Complete the bring up, troubleshoot, and final test.
7. Complete the documentation.

These steps are defined by their outcomes or completions. These are *milestones*, distinct from *tasks*. A task is the process you go through while you are taking action. A milestone is the point at which you have completed the task. Each step in the seven-step process is a task that ends in a milestone, the completion of the task.

When describing any plan, it is important to always pay attention to the difference between a *milestone*, which is a specific accomplishment that has been completed, and a *task* which is an activity in which you are engaged.

A milestone is "the schematic is completed." A task is "create the schematic."

1.8 Risk Management and Mitigation

The goal of any design project is to create a product that meets the performance goals at an acceptable cost, risk, and schedule.

There are often only a few right ways of completing a design on time and on budget, but many, many wrong ways which go over budget and over schedule.

Much of the design and implementation process is about navigating safe passage along the narrow path winding through the product development forest. One step off the correct path and your product may fall off the edge of a cliff into disaster, such as in **Figure 1.4**.

1.8 Risk Management and Mitigation   31

Figure 1.4. An example of a disaster: the Montparnasse derailment of 1895. Source is Wikipedia.

The plan of record (POR) should anticipate as many as possible of these potential problems which could cause your product to not work. For each problem or risk, a plan should be created to reduce or mitigate this risk. The combination of identifying potential risks and planning to reduce them is *risk management*.

Much of this book focuses on *risk reduction strategies*. We will anticipate the potential risk sites and create a mitigation plan to reduce these foreseeable, potential problems.

The basic approach we will take is:

Design more like Ralphie's Mom and less like a Colorado Bro.

1.9 Two Risk Management Design Strategies

There are two extreme approaches to risk management, which describe how to assess the risks in any endeavor. At the one extreme of thinking of every possible worst-case scenario is Ralphie's mom, in the 1983 movie, A [Christmas Story](). At the other extreme of not being concerned about any risk is a "Colorado Bro."

Ralphie's mom would always think of the worst-case scenarios, all the things that could go wrong, such as, "Be careful, you could put your eye out with that," referring to the possible consequences of using a BB gun.

A "Colorado Bro," a colloquial description of a young, risk-taking snowboarder, is less worried about what could go wrong and more interested in the thrill of "let's try it and see."

When a Colorado Bro encounters an opportunity to take a risky path, we have an expression in Colorado that, rather than being cautious, Coloradans will sometimes say, "Here, hold my beer."

These two approaches to life are illustrated in **Figure 1.5**.

Figure 1.5. Two design perspectives. Left: Design like Ralphie's mom: think of all the worst-case scenarios and avoid what you can. Right: A Colorado Bro's perspective is not afraid of taking on risk, "Here, hold my beer."

Designing more like Ralphie's mom means think of all the things that can go wrong, anticipate them, avoid them, or plan a risk reduction strategy. After all, the goal is first-time product success, on time, and within budget. A risk site is a feature or event that may prevent you from reaching your goal.

1.10 Master of Murphy's Law

Murphy's Law, as we use it today, is "If something can go wrong, it will." It was actually introduced by <u>Captain Edward Murphy</u> in 1949, shown in **Figure 1.6**. He was the lead engineer in the Air Force MX981 rocket sled project.

Figure 1.6 Air Force Captain Ed Murphy, the creator of Murphy's law, in 1949, source is Wikipedia.

This rocket sled was testing the impact on test pilots from the high G acceleration and deceleration they would experience in soon-to-be-introduced high-performance jet aircraft.

As part of the experiment, a test pilot was strapped onto a rocket sled on a railroad track, instrumented with sensors for G-forces, temperature, heart rate, blood pressure, and other physiological signals. These signals were routed to a tape recorder on the rocket sled.

After one particularly grueling day of testing, the recordings were reviewed and Captain Murphy found the entire day's recordings were blank. The excruciating pain the test pilots endured from the more than 40 Gs of acceleration and deceleration, leaving them bruised and even bloody, was wasted.

When Captain Murphy investigated the root cause, he found that the technician who wired up the sensors into the recording electronics had reversed each pair of leads. This led him to proclaim about the technician, "If there is any way to do it wrong, he'll find it."

This became an important principle for risk reduction: think of all the ways a project can go wrong and find a path to reduce the risk of each one happening before it happens.

The way to keep this law from screwing up your design is to anticipate the potential problems that could arise and put in place a plan to avoid the problems or check for them as early in the design process as possible.

How do you know what to anticipate as a problem that might arise? The more experience you have, the more problems you will have encountered and can anticipate for the next time. Learn from each problem.

You should not have to make all the mistakes possible yourself. Learn from the mistakes others have made or that you can imagine and add them to your list of mistakes to avoid.

Where can things go wrong? Consider in each phase of the design process:

- ✓ Could the part selection be wrong?
- ✓ Will the parts be available in the time you need them and at a cost within your budget?
- ✓ Could the schematic be wrong?
- ✓ Could the layout connectivity be wrong?
- ✓ Could there be too much noise generated by the layout?
- ✓ Could the wrong part be assembled at a location on the board?
- ✓ Could a package lead not be soldered to the circuit board?

- ✓ Could there be a problem with one of the components on the board?
- ✓ Could the testing introduce a measurement artifact, either a false positive or a false negative?
- ✓ Could the code be wrong?
- ✓ Would a customer have to refer to the manual all the time to use the product?

To reduce the chance of a problem arising:

- ✓ Think of all the specific details of each potential problem and design them out.
- ✓ Overdesign the product with extra margin to be robust if a problem does occur (maybe at higher cost, by buying insurance).
- ✓ Create a backup plan B, plan C, and even a plan D.
- ✓ Design more like Ralphie's mom.
- ✓ And most importantly:

> Don't add a new risk site unless you have a strong compelling reason to do so.

1.11 Proof of Concept

One of the most important risk sites in any product that uses a new and untested design is, will the design work? If you are combining together a new sensor, a new ADC, a new microcontroller, a new Bluetooth interface, a new H-bridge motor controller, a new DC motor, and a new articulating arm, how confident are you all the pieces will work together and that you will be able to move the arm with the accuracy you need to position the blocks with individual letters to spell the correct words?

Unless you are an expert in circuit design, hardware design, firmware design, and have confidence in the documentation for all the components, *do not assume your complex, intelligent product will work the very first time you complete the circuit board design, and assemble and integrate it with the rest of the system.* It is a much more realistic assumption that this complex product will NOT work the first time.

If you wait to test the entire system after you have designed and built the circuit board to connect all the components and acquired all the individual pieces of hardware, and have written the code for the microcontroller, and then find the system does not work, it may be so complex a system you will never find the problems, debug them, redesign the product, select new components and complete a second spin in the time available.

Instead, an efficient process is to divide and conquer. Break the entire product into functional blocks. Identify the critical elements or functional blocks that have the highest risk of design issues or unstable software control, and build these as prototypes as quickly as possible to verify the design will work.

Alternatively, if it does not work, you have a simpler platform from which to debug the hardware or software. You will get to an acceptable design faster with adequate time to modify the design or find other parts to make a successful prototype starting with a simpler system.

The purpose of these prototypes is to demonstrate the proof of concept (POC) and to work through and reduce the risk issues you identified as part of the plan of record. Will the part work at 3.3 V? Is the Bluetooth transmitter strong enough to reach the other side of the room? Will the SMPS you selected provide enough current for the H-bridge with acceptable noise?

The POC test vehicle does not have to look like your final product. It is a test platform. It could be a selection of commercial off-the-shelf (COTS) modules plugged together. It could be a solderless breadboard with components and hardware cobbled together to demonstrate all the components will play together.

1.11 Proof of Concept

An example of a simple POC used to test out a regulated high voltage generator circuit, a low-current pulse detector, and the code to measure pulses and output clicks and light flashes for a Geiger counter circuit is shown in **Figure 1.7**.

Figure 1.7 A simple proof of concept test vehicle used to demonstrate the key sensors, circuits, and firmware for a Geiger counter. This looks nothing like the final product.

To test out the firmware principles, the POC test vehicle should also include the microcontroller or processor you intend to use in the final product. The POC is a balance between simple enough to give you a high chance of success in testing out the minimum functioning, but complex enough to identify potential interferences your final product may encounter. The closer your POC looks to your final product, the earlier you can start software development.

The POC test vehicle should be completed as early in the design cycle as possible so that the very important risk site of "can the product be designed to work correctly" is reduced to an acceptably low level.

1.12 Practice Questions

1. What is the difference between a milestone and a task?
2. What does BoN stand for?
3. What are the seven steps to every project?
4. What is signal integrity?
5. What are the milestones to accomplish by the end of each step in the seven-step process?
6. What is risk management?
7. What is the primary focus of the POR?
8. What does it mean to design more like Ralphie's mom and less like a Colorado Bro?
9. What is Murphy's law?
10. Offer three examples of what you could do in a design to reduce risk that does not cost very much.
11. What is the purpose of a POC test vehicle?
12. What is the principle of divide and conquer?
13. What are three elements to the cost of a product?
14. What is the driving force for product design?

Chapter 2
PCB Technology

A printed circuit board (PCB) is the mechanical, electrical, and sometimes thermal platform on which components are assembled and interconnected, used in most products.

It is a proven, low-cost method of building electronic systems when electrical interconnect between components is required. While it evolved from the early days of interconnecting vacuum tubes, it is still used today, though in an advanced form, to interconnect the highest-performance integrated circuits.

Circuit boards are at the very heart of all electronic hardware products.

2.1 PCB, PWB, or PCA?

For historical reasons, we refer to the bare circuit board with no components attached as a printed circuit board (PCB) or a printed wiring board (PWB). After the components are attached, it is called a printed circuit assembly (PCA). Examples of PCBs and PCAs are shown in **Figure 2.1**.

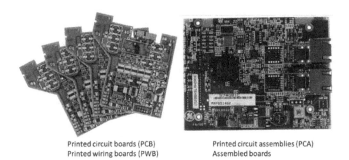

Printed circuit boards (PCB)
Printed wiring boards (PWB)

Printed circuit assemblies (PCA)
Assembled boards

Figure 2.1. The distinction between the terms printed circuit boards and printed circuit assemblies.

In this book, we will cover not just the design of the PCB interconnects but also the selection of the parts, their assembly, bring up, and test. These elements are part of the successful prototype development process.

2.2 Physical Design of a PCB

A printed circuit board is composed of alternating layers of conductors made of copper foil and dielectrics.

We define the number of layers in a circuit board in terms of how many metal layers are in the board regardless of how they are used. The simplest circuit board is composed of one conductor layer on a dielectric substrate. More commonly, most simple boards are two layers.

The board layers usually increase by two layers at a time. The next board stack-up is a 4-layer board, then, 6-layer board, and so on. This is due to the manufacturing process of laminating pairs of layers together to make multilayer boards.

Each dielectric layer is composed of a glass weave yarn impregnated with an epoxy dielectric. This is basically the structure of fiberglass. The most common dielectric material is called FR-4. This is "fire retardant" style #4.

Figure 2.2 shows a cross section of a signal line, copper planes, and the fiberglass dielectric between them. While there are three metal layers shown, this is just part of the board stack-up. On top of these are other metal layers.

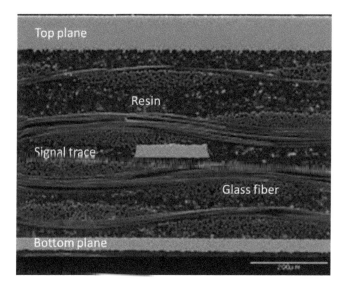

Figure 2.2. Cross section of a circuit board showing a copper plane on the top and bottom layers, with a signal layer between them. Source is Bogatin Enterprises.

But there is more to a circuit board than just the copper layers and the dielectrics between them.

2.3 Vias Technologies

With signals, power, and ground on multiple layers, electrical connections between layers have to be made with vias. There are three types of commonly used vias in a board, defined by how they are manufactured and what they connect. These three types of vias are illustrated in **Figure 2.3**.

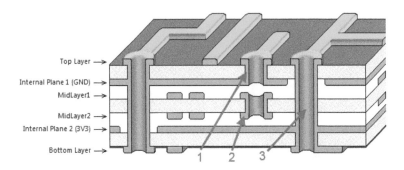

Figure 2.3. Examples of the three types of vias in a circuit board: (1) blind via, (2) buried via, and (3) through-hole via. Courtesy of Altium.

The most common type of via is a through-hole or plated through-hole (PTH) via. If you do not want metal on a specific layer to connect to the PTH via, add a clearance hole on that layer where the drill passes through the layer. This is done automatically in most routing tools based on net names. The clearance hole in the plane layer is sometimes referred to as an *antipad*.

To connect two inner layers, such as two ground planes, with a through-hole via, you would make sure there is no trace on the top or bottom layers where the drill passes through the board. The resulting through-hole via would only connect the inner layers.

The residual piece of plated through-hole barrel that sticks up or down from the last connected layer is called a stub. In very high frequency boards, with signals operating above 10 Gbps, the via stubs can be a serious signal integrity problem. However, for all boards operating at 5 Gbps and below, the via stubs are rarely a performance problem and can be ignored.

An example of a 4-layer board constructed with through-hole vias is illustrated in **Figure 2.4**.

2.3 Vias Technologies 43

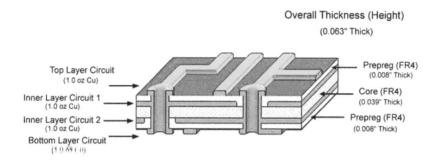

Figure 2.4 Example of a 4-layer board with through holes. Courtesy of Altium.

In this example, all the PTH vias drilled in the board go all the way through the board.

All the features and dimensions of the PTH are considered part of the via pad stack design. Unless you have a strong compelling reason otherwise, the features should be:

- ✓ Drill diameter 13 mils (this makes the finished, plated hole diameter 10.2 mils)
- ✓ Capture pad diameter around the via on any layer with a connection, 25 mils (6 mil copper annulus around the 13 mil drill diameter)
- ✓ Clearance annulus on a copper plane with no connection of 6 mils.

Part of the process of plating the via holes requires there be capture pads on the top and bottom layers of the via hole. The capture pad is to provide a larger target for the drill to hit to make up for registration tolerances. On layers on which there are no connections to be made by the via, there is no need for a capture pad. It is a good habit to remove these non-functional capture pads.

The annulus of the capture cannot be narrower than 6 mils for the lowest cost. This makes the outer diameter of the capture pad 13 mils + 6 mils + 6 mils = 25 mils.

With 6 mils as the narrowest clearance between copper features, the closest spacing between vias is 31 mils. A robust via-to-via pitch to use is 35 mils. This leaves 10 mils as the spacing between the edges of the via pads.

A buried via typically connects only between two adjacent layers. It is drilled and plated at the time when the 2-layer core is patterned. This extra drilling and plating step will add cost to the board.

The advantage of the buried via is that it does not block routing channels above and below it. This can be of value in very dense boards and may be worth paying the extra price. Most low-cost fab shops do not offer buried vias. For example, this type of via is NOT an option at JLCpcb.com.

A *blind via* is drilled from the top of the board to a controlled depth at the same time the through-hole vias are drilled. This makes a small, blind hole, from the top or bottom of the board, stopping on a lower layer.

After drilling, the inside of the hole has to be plated. This is difficult in a blind hole if the aspect ratio is larger than 1 to 1. It requires special chemistry to remove the air bubble in the hole and to get plating fluid to fill the hole and make contact to the exposed layers.

The value of a blind via is that a via can be placed inside a pad to make a low inductance contact to the layer below and it will not block routing channels on other layers.

If you were to add a through-hole via in a pad on which a component is soldered, the quality of the solder joint can be degraded. If there is a plated through-hole via inside a component pad, when the solder paste melts and reflows, the solder may wick down the plated hole by capillary action. This will steal or thieve some of the solder from the pad. This means there is the chance there may not be enough solder left on the pad to make a good solder joint to the component. This is why a through-hole via should never be placed in a pad if automated assembly will be used.

Blind vias are often used for via-in-pad applications for decoupling capacitors where low loop inductance connections are of high value. A blind via will not wick solder from a pad and will not cause an assembly problem. This is why blind vias are ok to use as via in pad connections.

Via in pad is not an option for many low-cost fab shops since it requires specialized plating chemistry and drilling.

> *Through-hole vias are the lowest cost and most common type of via. Unless you have a strong compelling reason otherwise, ALWAYS use through-hole vias and never any other type of via.*

2.4 Thermal and Thermal Relief Vias

There are really two purposes for a via:

- ✓ To conduct electrical current from one layer to another
- ✓ To conduct heat from one layer to another

The main purpose of a via is to provide electrical connectivity. These vias are referred to as just vias. But, when a via is used to conduct heat, it is referred to as a *thermal via*.

This should not be confused with a *thermal relief* via. A thermal relief via is a via used for electrical connectivity for which you want to limit the heat conduction.

If a via is connected to a plane, it will normally conduct heat very well to the plane. If this via is also connected to a component that would be soldered, such as with a pin inserted in the via hole, the plane may suck out so much heat as to make soldering very difficult.

For example, if a pin is to be inserted and soldered into a plated through-hole that also connects to a plane, the plane may suck so much heat from the pin as to make soldering the pin almost impossible. We need a way to reduce the thermal conductivity of a via but maintain the electrical conductivity.

How do you engineer a via to be both low electrical resistance but high thermal resistance? This is where a thermal isolation moat with a narrow drawbridge for electrical conduction comes in. An example of a thermal relief via is shown in **Figure 2.5**. The narrow tabs provide low electrical resistance, typically less than 0.5 mohms, but high thermal resistance.

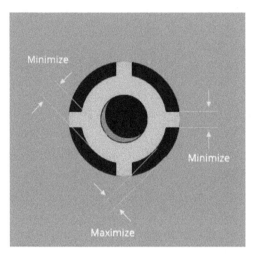

Figure 2.5 Example of a thermal relief via.

It is only when the via connects a soldered pin or is inside a solder pad used for the ground connection to a decoupling capacitor, for example, that a thermal relief via needed.

A thermal relief via will take up more room than a normal via. As a good habit, only use a thermal relief via when thermal isolation is also required.

When a via is between a signal trace and a plane and there are no solder pads adjacent to the via, there is no need for a thermal relief via. If the via is connected to a pad through a 6 mil wide trace, 20 mils long, there is no need for a thermal relief via. The trace acts as a thermal relief. If a via is between two signal traces on different layers, there is no need for a via with thermal relief.

Do not confuse a thermal relief via with a *thermal via*. A thermal via is used specifically to provide a low thermal resistance between a component and a plane. These vias usually connect the attach pad under a component such as a quad-flat-no lead (QFN) part and a plane underneath. An example of the footprint of a QFN showing the

2.4 Thermal and Thermal Relief Vias 47

thermal vias underneath the component footprint is shown in **Figure 2.6**.

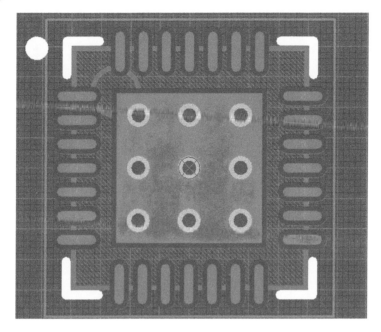

Figure 2.6 A small footprint QFN part with thermal vias in the center. Be sure to use thermal vias with a diameter at least 13 mils so there is no cost adder.

In many footprints for QFN parts, the thermal vias are by default 8 mil diameter or narrower in the footprint. If you use this footprint on your board, the fab shop may charge a premium price for the entire board just because it has narrow thermal vias. Be sure to change the size of the thermal vias and make them no narrower than 13 mils drill diameter.

Thermal vias act as a heat spreader for small components that may dissipate a lot of power. NEVER use thermal relief structures in a via being used as a thermal via to conduct the heat away from a component. It will defeat the entire purpose of a thermal via.

2.5 Other Layers

In a 2-layer board, the complete stack-up showing each layer a fab shop needs to return a completed circuit board is shown in **Figure 2.7**.

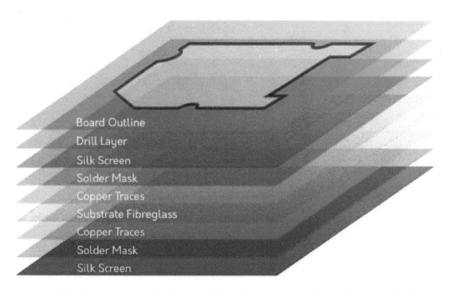

Figure 2.7. The complete stack-up of all the layers that must be delivered to the fab shop to return a completed circuit board.

The *outline layer* defines the shape of the board. It is this structure that is used to separate your specific circuit board in the panel in which it is built from other boards. An endmill will literally drill the board out from the rest of the panel. This is called "routing out" the board, not to be confused with the use of "routing" in regard to drawing traces on the board.

Many fab shops will not commit to fabricate your board without an outline layer. Without the outline layer, the fab shop does not know where your board ends and another customer's board begins in the large panel in which they are fabricated.

The *drill layer* has information about what hole sizes are used for each hole and their precise location, and is used by a precision CNC drilling machine to drill every hole in the board.

The *silk screen* is sometimes referred to as the overlay layer on both the top and bottom of the board. This is the printed text or numbers that identity each part and any other written reference information. It can also include images, logos, and pictures.

The *soldermask* is a layer of dielectric with small openings in it to reveal the underlying metal. Components will be soldered to pads exposed in the open areas of soldermask. The soldermask coating prevents the solder from flowing over the rest of the covered copper traces.

A very good review article that introduces many of the terms used in printed circuit boards can be found here, http://www.pcb.electrosoft-engineering.com/04-articles-custom-system-design-and-pcb/01-printed-circuit-board-concepts/printed-circuit-board-pcb-concepts.html

Download this short ebook here , https://www.asc-i.com/fundamentals-pcb-tech/download-e-book

2.6 The Soldermask Layer

On top of the copper layer is a soldermask that covers the entire surface of copper so it cannot be coated with solder, except for small windows made in the soldermask. These windows expose the pad on to which will be soldered the components.

The purpose of the soldermask is to keep solder only where it is supposed to be, the component attach pads, and not where it shouldn't be, the interconnect traces. If it weren't for the soldermask, solder would spread across the traces and not enough solder would stay on the pads to make a reliable solder joint for the component. The soldermask can also sometimes prevent solder bridging between closely spaced pads.

In addition to opening pads that isolate where solder will end up, the soldermask also protects the fine lines of interconnects from mechanical abuse.

The soldermask also is the mask used to expose the pads for a special surface finish. The soldermask has open windows exposing the bare copper. To these exposed pads, various surface finishes can be added. **Figure 2.8** is an example of the close-up of a board showing the soldermask defined openings as pads for soldering components.

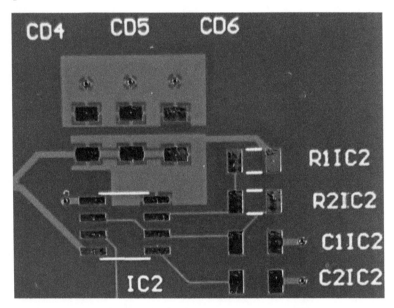

Figure 2.8 An example of the pads exposed by the soldermask layer to prevent solder from spreading over the board except where it is needed to attach a component. The exposed pads are then covered by HASL with leaded solder, in this example.

The color of a circuit board is really determined by the color selected for the outer layers, the soldermask, and the silk screen color. Most fab shops allow a variety of soldermask colors but may charge extra if it is not green soldermask and white silk screen.

Before you select navy blue soldermask and gold-colored silk screen, check the price in the quote from the fab shop. **Figure 2.9** is an example of some of the variety of colors available.

2.7 Surface Finishes

Figure 2.9 Examples of some of the soldermask colors available. The boards are all the same color. The only difference is the color of the soldermask laminated to the surface.

Try to resist (pun intended) the temptation of using a black, dark blue, or white soldermask. These tend to be opaque, which means you cannot easily see the traces underneath. This makes debugging more difficult. In the prototype phase, this is a soft error. When transferred to production, any color can be selected. The combination of the soldermask and silk screen color can provide a brand identification. Experiment with the combination you find most pleasing.

2.7 Surface Finishes

Bare copper easily oxidizes and will be difficult to solder if the board is left exposed for more than a week. Of course, it is only exposed copper on pads where components are attached for which we care about oxidation. Rather than using bare copper as the final surface metal on exposed pads, various metal coatings, called surface finish, are often added to make the exposed pads more easily soldered. The surface finish is a risk reduction.

Because a surface finish is so common on circuit boards, the option of not adding a surface finish and requesting bare copper is usually more expensive than adding a surface finish. This is because your board becomes a nonstandard board. You should have a strong compelling reason to pay extra for a bare copper finish.

The most common surface treatment is hot air solder level (HASL). For this finish, the final board with the bare copper exposed through the holes in the soldermask is dipped into molten solder, coating all exposed copper, and then pulled out through a hot air jet.

This is similar to the hot air hand dryer manufactured by Dyson. This leaves a thin layer of solder evenly coating the surface of every pad. The soldermask layer prevents solder from sticking to the board except in the openings in the soldermask, hence the name *soldermask*.

The solder is either leaded solder or lead-free solder. With a little solder flux added, pads with HASL of either type of solder can be easily soldered.

The lowest-cost HASL process currently at PCBway, and some other fab vendors, is HASL with leaded solder. This is because it is the lowest-cost soldering process and many panels are produced with leaded solder. Requesting HASL with lead-free solder is slightly more expensive because panels with lead-free solder use more expensive solder and a more expensive process and need to be tracked more carefully.

Before selecting leaded or lead-free HASL, check the price of the boards. For non-commercial applications, leaded solder can be used as easily as lead-free. If you have no other compelling reason, select the lowest-price process offered.

Even though your lab uses only lead-free solder, boards with leaded solder HASL are still allowed in most labs. Adding lead-free solder to a leaded solder HASL coated surface is not a problem. Either HASL solder will solder as easily using flux.

If you are selling your boards, to keep a RoHS (Restriction of Hazardous Substances) certification, it is worth the extra price to select lead-free HASL and qualify all of your components to be lead-free as well.

The second most common type of surface treatment is immersion tin. This is a very thin layer of tin applied in an electroless process by dunking (immersing) the board in a chemical bath. The board comes out with a very smooth and uniform surface. Like a HASL surface, an immersion tin surface can be easily soldered, and it is lead-free.

The third most common surface finish is electroless nickel immersion gold (ENIG). This process applies a thin layer of nickel to each pad, followed by a thin layer of gold. This surface is easiest to solder to and will last the longest when exposed to air. It is also as smooth and planar as the circuit board, since the coating is less than a fraction of a mil thick. It is a little more expensive.

Here is a useful article about surface finishes.

2.8 The Silk Screen

The final layer on top of the soldermask is the silk screen layer. This is often called the overlay layer. This is the printing with all the letters and words and symbols that is used to assist in assembly, bring up, test, and by the end user.

The finest pen width that can be printed, at no additional cost, is typically 6 mils. The pen width is also called the stroke. The character height should be at least 50 mils. Using characters smaller will be difficult to read. Use at least 50 mil characters unless you have a strong compelling reason otherwise. If you need to fit many small identifiers to a connector header, for example, you may need to use smaller character heights.

As a rough rule of thumb, keep the ratio of the character height to pen (or stroke) width about 6-10 to 1. This makes the characters readable without the letters smearing into a blob.

Depending on your design tool, a variety of fonts and point sizes can be used to create silk screen characters. Some fonts make very small, readable letters; others, not so much. This is a fun and creative area for experimentation.

At the very least, every board should have a unique board layout file ID name, your identification, and some rev or date code. A simple board name makes it easy to uniquely refer to the board. **Figure 2.10** is an example of the printing on a simple board.

54 *Practical Guide to Prototype Breadboard and PCB Design*

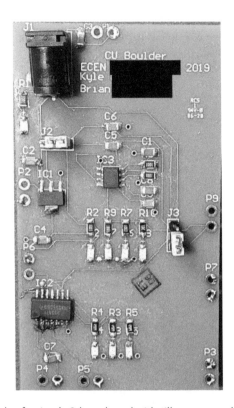

Figure 2.10 Example of a simple 2-layer board with silk screen markings to identify the board, the project owners, and reference designators to assist in assembly. Note that not all markings are helpful. This board was assembled by hand.

2.9 What the Fab Vendor Needs

When a fab shop builds the board, they need information about the metal patterns on each of these layers and where the holes will be drilled, their sizes, and whether they are plated or not plated with copper.

The set of files that have the pattern information on each layer is in the format called Gerber. These design files, and the drill files, are what needs to be supplied to a fab vendor to build the board. It is usually uploaded in a zipped file.

This specification and how to create the Gerber files are reviewed in chapter 24.

Watch this video to tour a commercial PCB fab and see how printed circuit boards are manufactured.

Watch this video to see how components are assembled to a circuit board and turned into printed circuit assemblies.

2.10 Practice Questions

1. What does the number of layers of a circuit board refer to?

2. What two layers are on top of the metal?

3. What is a good reason not to just use bare copper as the final metal finish on a board?

4. What does HASL stand for and why use it?

5. What does ENIG refer to?

6. If you plan to solder components to your board, what are two options for surface finish?

7. What is generally the lowest-cost surface finish?

8. What is a strong compelling reason to pay extra for a lead-free HASL finish?

9. What does the soldermask layer do?

10. What goes on top of the soldermask layer?

11. What are two pieces of information you might want to put in the silk screen layer on your board?

12. What is the minimum stroke to use in text on the overlay layer?

Chapter 3
Signal Integrity and Interconnects

The primary purpose of the PCB is to provide the electrical connections between all the terminals of all the components that need to be connected together to function. We describe this as *connectivity*.

When the electrical properties of the interconnects between terminals and between nets do not affect the signal quality in any way, we refer to these interconnects as transparent.

But, when the electrical properties of the interconnects, described by their parasitics, do affect the noise generated at component terminals, we refer to the interconnects as *not* transparent. When interconnects are not transparent, they will *only* degrade signals by introducing noise.

> *The interconnects in the PCB can never improve the performance of the product above and beyond what the individual components are capable of. Once connectivity is established, the only thing the interconnects can do is add noise to the signals and power delivery and potentially contribute to electromagnetic interference (EMI) test failures such as from Federal Communications Commission (FCC) certification.*

For more details on this general topic, view the keynote I presented at Altium live: https://resources.altium.com/p/the-value-of-the-white-space-pcb-design-eric-bogatin.

3.1 Transparent Interconnects

If the noise contributed by the interconnects is below a threshold that will impact the product in any way, we refer to the interconnects as *transparent*. Their electrical properties do not influence the product performance.

We describe the electrical connections of all components as they are designed to be connected as *design for connectivity* (DFC). This means all the nodes and terminals are connected correctly and terminals that are not supposed to be electrically connected are not connected.

All terminals or pins connected together by wires are considered one *net*. The net is defined by which components' pins are all connected together. The collection of all the nets in the circuit board and which terminals are connected to each net is referred to as the *net list* and is the database description of the circuit board.

If the product *works* in a solderless breadboard with no regard for using best design practices and wires going every which way, the interconnects are probably transparent. If it works in a solderless breadboard, it will very likely work in a printed circuit board with no special signal integrity considerations for the layout of the interconnects. In this case, it won't matter how the board is designed, other than for connectivity. The interconnects are transparent.

An example of a product with interconnects that are transparent is shown in **Figure 3.1**.

Figure 3.1 The product on the left works with jumper wires. It doesn't matter how the circuit board is designed. If the connectivity is correct, it will work.

3.1 Transparent Interconnects

In products where the interconnects are transparent, the layout of the wires and placement of the parts may add noise, but it will be within an acceptable level that will not influence the product.

It is unfortunate that many design guidelines are created based on features that were used in a design in which the product worked and it was assumed the features were required to make the product work. If the interconnects were really transparent in the application, the product worked in spite of the features added. For example, this is the origin of the recommendation for a ground pour on signal layers or the use of rounded corners for traces.

While there are many electrical noise problems that can arise, most scale inversely with the rise time of the signal and linearly with the length of the interconnects. The shorter the rise time, the more the noise. The longer the interconnects, the more the noise.

Fundamentally, this is because many sources of noise scale with a dI/dt or dV/dt and the interconnect electrical properties that create the noise, such as its loop inductance or capacitance between signal interconnects, scale with the length of the interconnect.

This is why some interconnects can look transparent: for the rise time of the signals with the lengths used in the product, the interconnects are transparent. An example of the noise in an interconnect measured by a scope for two different length interconnects, is shown in **Figure 3.2**.

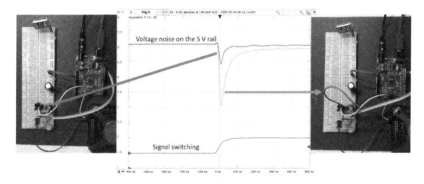

Figure 3.2 Using a short wire routing power to a switching transistor results in a reduction of voltage noise on the power rail from a 2 V drop to only a 0.5 V drop.

In this example, the voltage on the 5 V power rail is measured on the collector of a transistor. Nominally, it is 5 V. When the transistor switches a 400 mA current, this dI/dt creates a voltage drop on the 5 V rail, seen as a dip on the collector.

When the wires connecting the transistor to the 5 V rail are short, the noise dip is only 0.5 V. When the interconnect wire connecting the transistor to the 5 V rail is longer, the voltage drop noise is as large as 2 V, out of a 5 V rail.

An interconnect of some length that might have been transparent with one rise time may not be with a shorter rise time. At a fixed rise time, shorter length interconnects will usually be more transparent.

3.2 When Interconnects are NOT Transparent

When are interconnects *not* transparent? The answer is, as is the most common answer for all engineering questions, "it depends."

But it is not enough to say it depends. For engineers, the very next sentence must state what it depends on, and, using simple estimates, provide some metric of under what conditions upon which it depends. This is the essence of engineering.

Of course, all the details of the circuits and the nature of the interconnects affects this boundary of when interconnects are transparent, but we can make a very rough estimate.

> *An important principle in engineering we will use over and over again is "Sometimes an OK answer NOW! is more important than a good answer later."*

Even though the real problem is complicated and depends on many factors, sometimes we just want to get a rough idea of when an effect might be important. This is when simplifying assumptions and rules of thumb are incredibly valuable in providing a rough estimate, quickly.

3.2 When Interconnects are NOT Transparent

The two most important noise sources that will usually become large enough to be a concern are each a type of *switching noise*, noise that arises when one or more signals change state and their currents turn off or on.

One type of switching noise is found on the power rail and is caused by the changing power rail currents passing through the power lead inductance between the IC power pins and the voltage regulator module (VRM) or the nearest decoupling capacitor. The VRM is the component that provides the regulated voltage to the power rail on the board. This type of noise is specifically called *power rail collapse*.

The second type of switching noise is found on signal paths and is generated due to changing return currents passing through the higher inductance of a return path that is not a wide, continuous plane. It increases with more signals sharing the same return conductor and switching simultaneously. We refer to this sort of noise as *ground bounce*.

This is noise generated by multiple signals changing their state with all of their return currents passing through the same, narrow conductor. This happens in IC packages, in connectors, in some multiconductor cables, in solderless breadboard interconnects, and in circuit boards with poorly designed return paths.

It is fundamentally due to the changing or switching currents passing through the shared inductance of a common return path. An example of ground bounce noise on a victim line with one, two, three, and then four I/Os switching simultaneously through traces on a board with a shared return trace is shown in **Figure 3.3**.

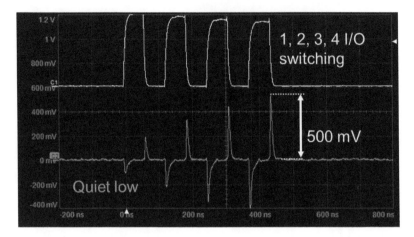

Figure 3.3 Measured 5 V signal on the aggressor trace (top) and the measured induced noise on the victim trace (bottom) when an increasing number of aggressor signals switch. This is a form of cross talk, specifically ground bounce or simultaneous switching noise.

Using a few simplifying assumptions, we can estimate when ground bounce noise may be a problem.

The voltage noise across the inductance of an interconnect is:

$$\Delta V = (L_{Len} \times Len) \times \frac{\Delta I}{RT}$$

Where

ΔV = the voltage noise generated, mV

L_{Len} = the inductance per length of the interconnect, nH/in

Len = the length of the interconnect, in

ΔI = the change in current or the transient current, mA

RT = the rise time of the changing current, nsec

If we call a transparent interconnect when the induced voltage noise is below some maximum acceptable level, then the relationship between the longest length that is still transparent is,

3.2 When Interconnects are NOT Transparent

$$\text{Len}_{transparent} < \frac{1}{L_{Len}} \times \frac{\Delta V_{max}}{\Delta I} \times RT$$

Now we can put in some numbers.

For simple interconnect wires, the inductance per length is about 20 nH/inch. If we say voltage noise less than 50 mV is insignificant when there is 50 mA of current change, the criterion of transparent interconnect becomes.

$$\text{Len}_{transparent}[\text{in}] < \frac{1}{20[\text{nH/in}]} \times \frac{50\text{mV}}{50\text{mA}} \times RT[\text{n sec}]$$

$$\text{Len}_{transparent}[\text{in}] < 0.05 \times RT[\text{n sec}]$$

This rule of thumb says that if the shortest rise time of a signal in a circuit is 100 nsec, interconnects shorter than 5 inches will look transparent. They will have negligible noise, and how we route the signals and return paths in the solderless breadboard or the circuit board will have little impact on the performance of the circuits. This criterion defines two regions of design space, when interconnects are transparent and when the design of the interconnects matter. This is illustrated in **Figure 3.4**.

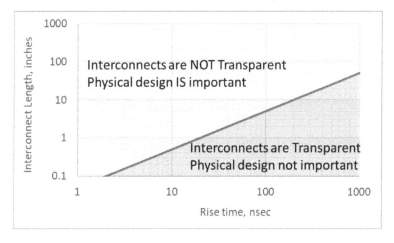

Figure 3.4 Mapping the design space where interconnects are transparent (lower triangle) and interconnect design is not important for this set of assumptions.

Most solderless breadboards have wires less than about 10 inches long. This suggests that as long as signal rise times are longer than about 200 nsec, the interconnects in solderless breadboard circuits will be transparent. This is a very useful rule of thumb.

But this is a very conservative estimate. If the current changing in a circuit is not 50 mA, but goes up to 500 mA, the length condition for transparent interconnects is reduced by a factor of 10. For a rise time of 200 nsec, any interconnect shorter than only 2 inches would be transparent.

In addition, even if there is significant switching noise, if it occurs at a time at which the circuit is not sensitive, it may not be important.

This is why so many circuits work when implemented in a solderless breadboard or in a circuit board despite following no signal integrity design guidelines or following inappropriate guidelines: the interconnects are transparent, or the signals are insensitive to the noise levels generated.

By following best design practices using solderless breadboard interconnects, it may be possible to extend their useful range into rise times as short as 2 nsec if interconnect lengths can be kept short. **Figure 3.5** shows the measured signal from a small micro controller module plugged into a solderless breadboard with a measured 10-90 rise time as short as 1.95 nsec using a short interconnect.

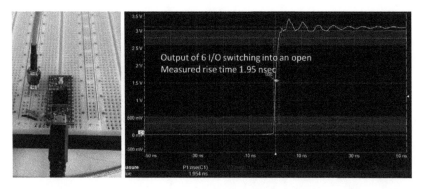

Figure 3.5 A Teensy 4.0 module in a solderless breadboard measured with a high bandwidth probe and 8 GHz bandwidth Teledyne LeCroy WavePro HD scope. The measured rise time is 2 nsec.

The slight ringing noise evident on the scope trace is actually due to cross talk noise on the IC package and small printed circuit module, not from the solderless breadboard.

As rise times decrease, the noise levels will increase and there will absolutely come a short-enough rise time or long-enough interconnect when the switching noise will be large enough to be a concern. This is when the design of the interconnects to reduce the inductances and other noise sources in interconnects will become important.

Because precision-engineered interconnects can be designed and fabricated in printed circuit boards to control their electrical properties, it will always be possible to build circuit boards with less noise-producing interconnects than the best case in solderless breadboards. This is why the highest-performance (shortest rise times, longest interconnect length) circuits will always be constructed in printed circuit boards.

This analysis points out how important it is to know what the rise times of your signals are. If you don't have any idea what your rise time is, it is difficult to evaluate when the interconnects will be transparent and when interconnect design matters.

3.3 Where Signal Integrity Lives

Every design, implemented as either a solderless breadboard or a printed circuit board, starts with a block diagram or functional diagram that is quickly turned into a schematic.

The schematic is the collection of the symbols representing each component used in the circuit, and how they are connected together. This is a graphical representation of the database that lists each component, the terminals on the ends of each component, and how the terminals connect together.

When we draw the connections between the terminals of the components on the schematic page, we are graphically defining the connectivity for the net list. The connections we draw, either as discrete wires or as labels called net names, which define to which net each terminal connects, define the connectivity. An example of a schematic for wiring up an Atmega 328 microcontroller is shown in **Figure 3.6**.

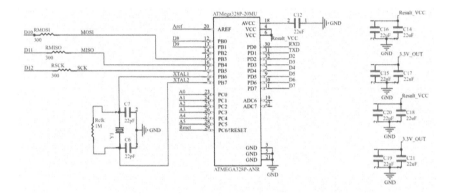

Figure 3.6 An example of a schematic showing the components and how their terminals are connected together.

The wires we draw in the schematic are perfectly transparent. They have no electrical properties other than 0 ohm resistance and 0 time delay. As Shawn Hailey, one of the founders of MetaSoft (acquired by Synopsis) and a cocreator of HSPICE, is fond of saying, "Wires in a schematic are Tachyonic superconductors."

This means the wires drawn in a schematic will never create noise of any sort. If the connectivity is correct, they will never contribute to electrical problems. They are perfectly transparent. A schematic says absolutely nothing about signal integrity.

The problems with the interconnect arise when the schematic is translated into the physical features of traces on a circuit board in the layout process.

Noise is always due to an *aggressor* signal on an aggressor interconnect generating noise that appears on a *victim* interconnect. When the noise distorts the aggressor signal on the aggressor interconnect due to its own interconnect, we refer to this as *self-*

aggression noise. When the noise from the aggressor is generated on another interconnect, we refer to this as *mutual-aggression* noise.

Self-aggression noise is generated on the same net as where the aggressor signal propagates. Self-aggression noise on a net is generated by the physical implementation of the wires that make up the net.

Mutual-aggression noise is transmitted through the space between the aggressor and victim nets, literally through the insulating dielectrics or air between the conductors. It is transmitted by the electric and magnetic fields of the aggressor signal from the aggressor interconnect, coupling to the victim interconnect through the space (air) between them.

Even though there is no DC current path between the aggressor and victim nets, currents can flow literally through the air or insulation by either displacement current (the dV/dt across capacitance) or through induced currents between a mutual inductance. These currents flow because of the changing electric and magnetic fields between the aggressor and victim loops.

We call the field lines that span between the aggressor and victim signal-return loops *fringe fields* because they extend beyond just the local vicinity of the aggressor conductors. An example of these fringe electric and magnetic fields from an aggressor signal-return path to a victim signal-return path is illustrated in **Figure 3.7**.

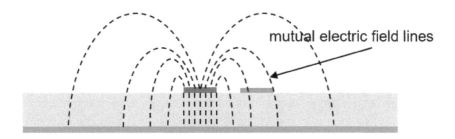

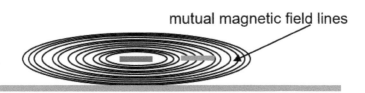

Figure 3.7 A cross-section view of an aggressor and victim signal conductor and their return path, showing their fringe electric and magnetic field lines that create cross talk. The shared field lines are referred to as mutual field lines.

> *Signal integrity problems lurk in the invisible spaces in a schematic: in the wires and in the white space of the schematic. They are only brought to life in the layout.*

It is an unfortunate fact that once connectivity is defined by the wires, the performance of the circuit will only be degraded by the interconnects.

The electrical properties of the interconnects can never improve the performance of the circuit over what the components themselves are capable. The interconnects can only degrade performance and screw things up.

When we design interconnects, one of our goals is to engineer them to limit how much they screw up the signals and keep the noise they generate below an acceptable level.

There are some best design features we can implement, which are free, that will always contribute to less noise. If they really are free to implement (no cost), and they reduce noise (add value), they offer a very good ROI and should always be implemented. We call these *habits*.

Practice including these features in your boards even if you don't need them in your current design for it to work. They may not be necessary in your current design, but guaranteed, there will be a future design where these design guidelines are important. Experience incorporating these habits in all your designs will make you a better designer.

For example, the most important feature we can add to a board to reduce one type of switching noise is to use a continuous ground plane, or return path, on a copper layer adjacent to the signal traces.

We can use a signal trace in the vicinity of the aggressor trace on which we purposefully keep the driven signal at a low level. This trace becomes a sense line for any noise. If there is no signal on the trace, but we measure a voltage, this voltage must be due only to noise. We call this sort of sense line a *quiet* line. It is a very important way of measuring just the noise that would appear on a trace. Of course, when a signal is imposed on the trace, the voltage measured would be the signal plus the noise.

Figure 3.8 shows the measured noise on a nearby victim trace when signals propagate on an aggressor interconnect over a ground plane and in a separate region of the circuit board where there is no ground plane. The measured noise on a similar quiet trace is reduced by a factor of 30 using a continuous ground plane underneath, acting as the return path in this example.

70 *Practical Guide to Prototype Breadboard and PCB Design*

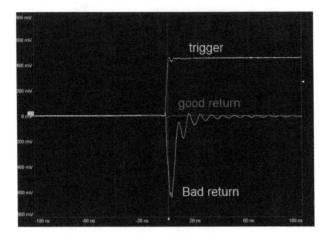

Figure 3.8 Measured noise on two different quiet lines when an aggressor line switches in two identical circuits with different layouts. When there is a continuous return plane, the noise is less than 20 mV. When there is no continuous return plane, the noise is 600 mV.

The schematics for these two circuits are identical. Their only difference is in how the layout was implemented. One circuit has a continuous return plane under all the signal traces while the other layout just has a trace as the return path.

This specific noise is generated because of a changing current in the aggressor interconnect passing through the mutual inductance shared with the victim interconnect. It is an example of *switching noise*. Since it arises due to a screwed-up return path, it is also a form of *ground bounce*. And because it scales with the number of signals switching simultaneously with their return currents sharing this screw-up return path it is also a form of *simultaneous switching* noise.

> *Watch this video and I will walk you through how this ground bounce noise is measured.*

3.4 Six Categories of Electrical Noise

The interconnects can only decrease the performance of a circuit, never improve it. This means that after connectivity, their only contribution in a system is to add noise. We group these noise sources into three general categories:

Signal integrity: noise problems associated with the *signals* that propagate through the interconnects.

Power integrity: noise problems associated with the delivery of the nominally constant *DC voltage* to each active device, usually caused by *power and return* currents.

EMI: noise problems associated with the interaction of the external world by *radiated emissions*, either from the circuit to the external world (emissions) or from the external world onto the circuit (susceptibility). The term electromagnetic *interference* (EMI) is used when describing the problems, and electromagnetic *compatibility* (EMC) is used when describing the solutions.

While there are clear and succinct problems that arise in each of these areas, some problems overlap, and these three general areas of noise should be considered holistically. An illustration of this holistic approach to noise problems caused by interconnects is shown in **Figure 3.9**.

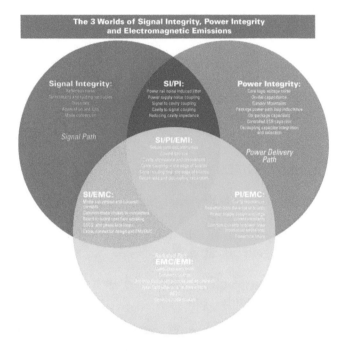

Figure 3.9 Three families of noise problems require a holistic approach. Courtesy of *Signal Integrity Journal*.

Every type of noise is a response on a *victim* interconnect from a stimulus-signal on an *aggressor* interconnect. For every type of noise, there are always four elements:

- ✓ The *stimulus*: the aggressor voltage or current signal.
- ✓ The aggressor's *interconnect* on which the aggressor signal appears.
- ✓ The *response*: the induced noise voltage or current as the unwanted signal on the victim interconnect.
- ✓ The victim's *interconnect* on which the victim noise appears.

When the victim's interconnect is also the aggressor's interconnect, we call this *self-aggression* noise.

When a signal propagates from a transmitter (TX) on an interconnect, it may suffer reflections that head back in the direction

3.4 Six Categories of Electrical Noise

toward the TX. The reflections are scaled versions of the signal, generated, propagating in the opposite direction, that can in turn re-reflect and bounce around the interconnect and interfere with the incident signal at the receiver (RX). This is a form of self-aggression noise. **Figure 3.10** shows an example of ringing noise at the RX due to excessive self-aggression noise in the form of reflections.

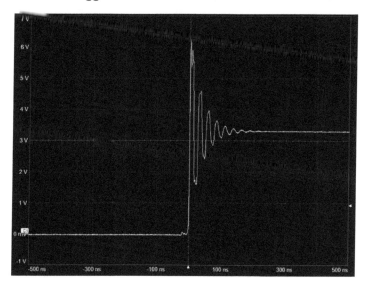

Figure 3.10 Measured voltage on a receiver showing the ringing noise due to multiple reflections on the interconnect. This is a type of self-aggression noise.

When the victim is a different net than the aggressor, we call this *mutual-aggression* noise. This is the general class of noise called cross talk. The signal on one net creates noise on another net, the victim.

The aggressor net does not have to have any direct electrical DC connection to the victim net to induce noise. Electric and magnetic fields coupling through the air or dielectric between the two nets can induce noise on the victim net. This is a form of *near-field* coupling.

This makes mutual-aggression noise sometimes difficult to trace back to its source, since the coupling is literally through the space between the conductors and is difficult to see directly. This is also why we say signal integrity lives in the white space of the schematic.

Figure 3.11 shows an example of the noise on one net due to the signal on another net. Any signal that would also be on the victim net would see its signal AND the noise added in. If the noise is large enough, it can interfere with or degrade the signal on the victim line.

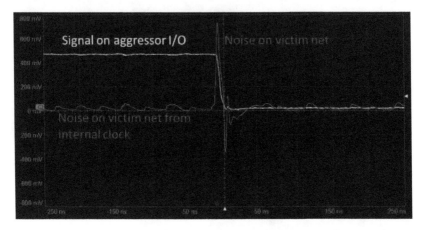

Figure 3.11 Measured switching noise on victim net, more than 1 V peak to peak, from a 5 V aggressor signal switching.

In each of the three categories of SI, PI, and EMI, there are noise sources that result in self-aggression and mutual-aggression noise. This combination is six different types of noise.

Using this taxonomy to describe the various noise sources will help us describe the specific type of noise and identify its root cause. Understanding the root cause of a noise source is the first step in reducing it. As we go through the various noise sources, we will identify their root cause and how we engineer the interconnects to reduce this problem.

3.5 Families of SI/PI/EMI Problems

The fastest way of fixing a problem is by identifying its root cause. Then, fix the problem based on adjusting the features that contribute to the root cause.

3.5 Families of SI/PI/EMI Problems

When we look at all the various types of electrical noise that are possible, we find there are some noise types that are most common and occur over and over again in electronic products, unless special care is taken.

There are six commonly encountered sources of noise that we can group into families based on their root causes. Some of these are considered in more detail throughout this book.

Reflection noise: appears as ringing noise at the receiver and is due to reflections from impedance changes. It only appears when interconnects are electrically long compared to the rise time of signals. Not all noise that appears as ringing is due to reflection noise. If the victim interconnect is electrically short, it will not show reflection-based ringing noise. This is a form of signal integrity self-aggression noise.

Cross talk: is due to the fringe electric and magnetic fields from the aggressor interconnect coupling to a victim interconnect. In electrically long interconnects, cross talk on the victim line will appear with a slightly different signature at the end near the aggressor TX or at the end far from the aggressor TX. In electrically short interconnects, the cross-talk noise will appear with the same signature everywhere on the interconnect. This is a form of signal integrity mutual-aggression noise.

Ground bounce: is a special case of cross talk and a type of switching noise when there is coupling through a common ground or return path shared by multiple signal lines. This is the most common type of cross talk found in electrically long and electrically short conductors. It can appear as inductively coupled noise and resistively coupled noise. This is a form of signal integrity mutual-aggression noise. Since it scales with the number of switching signals that switch simultaneously, it is also called simultaneous switching noise (SSN).

Rise time degradation is associated with very short rise time signals on long interconnects. The rise time of the signal is increased by the time it gets to the receiver due to frequency-dependent attenuation of the higher-frequency components of the signal. It is only an issue with very short rise times or long interconnects. As a rough rule of thumb, it will only be apparent when the interconnect length in inches > 100 x rise time in nsec. This is a form of signal integrity self-aggression noise. It is the dominant source of noise in high-speed serial links such as PCIe busses.

Power rail collapse: occurs when the voltage on the power rail feeding a device drops due to the transient current passing through the impedance of the power and ground return interconnects, referred to as the power distribution network (PDN). This is dominated by the inductance of the power path interconnects. It is reduced by placing decoupling capacitors in close proximity to the IC to reduce the series inductance. This is a form of switching noise and power integrity self-aggression noise.

Electromagnetic interference: occurs when some of the near field emissions from signals on interconnects turn into far field emissions. These emissions can sometimes result in very long-range cross talk and fail FCC certification tests. While near field emissions arise from all signals and their return paths, it is only when the separations between the signals and their returns are large, such as when return paths are not in close proximity to signal paths, that their near fields also extend into the far field. This is a form of EMI mutual-aggression noise.

By understanding the principles of how to turn physical interconnects into equivalent electrical circuit models and how the properties of these circuit models, driven by signals, cause electromagnetic noise, we can analyze the root cause of these families of noise, engineer a solution to minimize them, and have confidence before we build the final product, they will have been reduced to an acceptable level.

3.6 In Principle and In Practice

No matter what we do, noise from each of these six categories of problems will *ALWAYS* be present in *EVERY* design. You can never eliminate any of these noise sources, *in principle*.

In principle means that an effect is possible under some set of conditions. However, it says nothing about the magnitude of the

3.6 In Principle and In Practice

effect or the probability of it occurring or the frequency at which it actually occurs.

In principle, a 1 mA signal with a rise time of 1 second can create switching noise on a victim line. In principle. However, *in practice*, the magnitude in any reasonable system would be on the order of 1 mA/1 sec x 10 nH = 0.01 nV, a level well below any influence in circuit performance.

In principle, a right-angle corner on a signal trace will cause reflections and affect the signal. Depending on the line width, the impact of a 90-degree corner is to add about 50 fF of excess capacitance to the trace. In practice, for signals with rise time longer than about 50 psec, this is an insignificant effect and is hardly noticeable.

Whenever you use the term *in principle*, it must be qualified, by putting in the numbers, with the magnitude of the expected effect. Then the effect can be evaluated as important or insignificant. This is the difference between what can happen in principle and what may happen in practice.

> *The way we separate what can happen in principle with what may happen in practice is by putting in the numbers, and estimating the magnitude of the effect using simple rules of thumb, approximations, or numerical simulations.*

Many behaviors that can happen in principle can be reduced to either an acceptable level or even to a level below which they cannot be measured, in practice. Always be aware of what can happen in principle but get in the habit of putting in the numbers, and estimating the magnitude of the effect, to determine if this may be a problem in practice.

In addition to engineering the interconnects to reduce these noise sources, the techniques to measure them are also important. Even if these noise sources do not prevent the product from "working," measuring their noise level gives us an estimate of how far away they are from causing a problem. This is referred to as the *performance or design* margin.

Knowing the performance margin in a product provides a level of confidence in the risk that noise will not cause a problem and is direct feedback on the impact your design features have on the creation of noise.

Often, the evidence cited to support a specific design decision is that with this design feature, the product worked, so it must be the right thing to do.

> *Do not confuse coincidence with correlation. Just because a feature is added to your design and your design works does not mean that feature was instrumental in assuring your design works.*

For example, the very common practice of using three different capacitor values on a power pin of 0.1 uF, 1 uF, and 10 uF is often justified because the product worked when this was implemented. **Figure 3.12** is an example of a schematic taken from a contemporary, 2020, microcontroller design using this combination of decoupling capacitors.

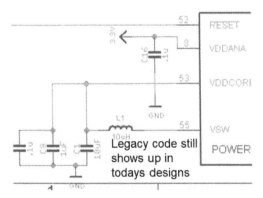

Figure 3.12 An example of a contemporary schematic design showing the requirement for three different capacitor values on a power rail.

In many cases, the product probably could have "worked" with no capacitors, or with three different values, or with three capacitors all the same value, or with 1 capacitor. The product worked *in spite of* the design decisions for the decoupling capacitors.

The most important design condition for the decoupling capacitors is low loop inductance between the IC pads and the decoupling capacitor. This question is addressed in this online article: https://www.signalintegrityjournal.com/articles/1589-the-myth-of-three-capacitor-values.

While it is possible to make design decisions based on the process of "build it and test it," a much more effective strategy is to:

1. Understand the underlying essential electrical principles

2. Apply these principles to the circuit design and interconnect electrical properties.

3. Identify the root cause of the noise.

4. Engineer a solution that minimizes the noise based on the root cause.

5. Analyze the physical interconnects in terms of their equivalent electrical circuit models.

6. Estimate the expected noise in this new design using rules of thumb, analytical approximations, and numerical simulation tools (virtual prototype).

7. Build an evaluation vehicle (real prototype) to measure noise and confirm the expected noise based on the estimates in order to gain confidence in the simulation analysis.

When you apply this process to reducing signal integrity noise from your interconnect design, you rachet up the learning curve and have confidence your design decisions will result in acceptable performance in your current design *and* your next designs.

3.7 Net Classes and Interconnect Problems

The schematic says nothing about signal integrity. But, if we have an idea of the types of problems specific nets might encounter due to their signal rise times and estimated interconnect lengths, we can pass along important design guidelines to the layout.

One way of doing this is by assigning net classes to the nets in the schematic for which we expect specific noise problems, which carry with them an implied set of design guidelines to follow to reduce these problems.

These net classes basically define the problems that might arise and the interconnect design guideline solutions to reduce them.

All nets should always be designed for the default case to reduce switching noise in the form of ground bounce and rail collapse. Following the important design guidelines is usually free and is a good habit to follow.

The *default signal trace* net class should be designed with the minimal line widths, typically 6 mil wide and via dimensions the fab vendor can manufacture at no cost adder, typically 13 mil drill hole size. These paths can typically handle up to 1 A of DC current. They should also be designed with continuous return paths to reduce ground bounce switching noise. This is the default net class for every signal net for which transmission line effects are not important.

The *default power path* net class applies to power traces that will carry less than 3 A of current. These should be designed with 20 mil wide traces. Decoupling capacitors should be placed in close proximity to the IC they are decoupling to reduce rail collapse switching noise.

The *high current (> 3 A) power* net class should be designed with wider traces to keep the trace temperature-rise below an acceptable value like 20 decC.

The *controlled impedance* net class is for nets that meet the default conditions with the added condition of the signal rise time being short enough and interconnect length long enough where reflection noise may be an issue. The reflection noise is reduced by engineering interconnects as controlled impedance at a specific target impedance and implementing a termination strategy for signals either on-board or off-board.

The *routing topology sensitive* net class comprises nets in which, in addition to the features of the controlled impedance net class, reflection noise can arise because of the trace routing geometry, such as branches in the net. The routing topology, controlled impedance, and termination strategy are engineered to minimize the noise on these nets.

The *cross talk sensitive* net class comprises nets where, in addition to the routing topology net class concerns, the spacing and routing between nets will influence the mutual-aggression noise coupled onto victim nets. All of the above considerations are also important, in addition to the relative spacings.

The *differential pair* net class has all of the issues described above for single-ended signals and the additional concern for length matching of the two lines that make up a differential pair.

The *high-speed serial* net class, generally also differential pairs, has all the issues of the differential pair net class and the added concern for the conductor and dielectric losses in the interconnect. These interconnect features will increase the rise time of signals and collapse the eye. This net class generally applies to nets carrying signals with data rates above 1 Gbps and interconnects longer than about 10 inches.

The *discontinuity sensitive* net class comprises nets for which small regions of the interconnect are not a controlled impedance or at a different impedance than the rest of the net and influence the self-aggression reflection noise. These are nets that contain structures such as termination resistors, DC blocking capacitors, vias, and connectors or test points. Generally, if reasonable care is taken to design controlled impedance interconnects, discontinuities will play a dominate role with signal bandwidths > 1 GHz.

When the schematic is developed, associating specific nets based on their expected rise time and possible interconnect lengths will give a heads up to the layout design team as to what special design considerations should be taken into account to reduce the expected noise problems.

82 *Practical Guide to Prototype Breadboard and PCB Design*

The design guidelines for each of the different net classes are the basis for most of the field of signal integrity.

When interconnects are not designed with these problems in mind or steps are not taken to minimize these problems, the signals at the receivers can be grossly distorted. **Figure 3.13** shows what can result when these net classes and the design guidelines to follow for each net class are not taken into account.

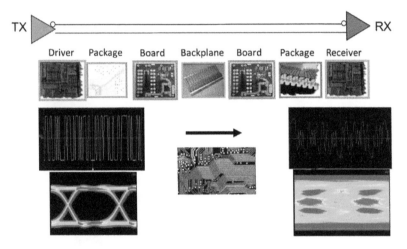

Figure 3.13 An example of the signal coming from a transmitter (TX), and how the interconnect can distort the signal by the time it gets to the receiver (RX). Engineering each net class for acceptable performance will open up the eye at the receiver.

3.8 Design for Performance

Once connectivity is established, to avoid the anticipated noise problems, we will need to design for performance (DFP). The focus of this book is establishing the design guidelines for the default net class: the guidelines every signal and power interconnect should follow to reduce the switching noise from ground bounce and power rail collapse. As an added benefit, following these guidelines will also reduce most sources of EMI.

We break these design-for-performance guidelines into four areas:

- *DFP: at DC*

3.8 Design for Performance

- *DFP: signal paths*
- *DFP: power delivery*
- *DFP: EMI*

The best design practices for DFP are covered in detail in later chapters. However, some design practices are so important, they will be stated repeatedly so they become ingrained as habits.

When a DFP feature will reduce the potential noise and not add to the cost of the product, it is a best design practice and should be adopted as a habit.

DFP design guidelines will reduce the potential noise. When implementing them is free and does not add cost, we should always follow them regardless of whether the interconnects are transparent or not. They should become habits, unless we have a *strong compelling reason* otherwise.

The six most important habits for the default net class conditions for DFP to reduce noise and risk are:

1. Use as narrow a trace and as small a via as your fab shop allows to enable higher interconnect density and ease of routing. Even a low-end shop can do 6 mil wide traces and 13 mil drilled holes. This offers the highest interconnect density at no additional cost. Any smaller than this and not all fab shops can implement the features at no additional cost.

2. To reduce ground bounce, route a continuous ground-return plane under all signal traces. If you need to use a cross under to the ground plane for routing, keep the cross-under as short as practical. If you have multiple signals passing across a gap in the return plane, which you can't keep shorter than ½ inch long, add shorting return straps across the gap.

3. Route adjacent signal traces as far apart as practical to reduce cross talk, even if you have a continuous return plane on the bottom layer.

4. In power paths that carry less than 3 A, use a 20 mil wide trace. This way, at a glance, you can identify which traces are power paths and which are signal paths.

5. Use as small a loop inductance as practical in the path from the power pins of an IC to the decoupling capacitors for that IC. This means short, wide power paths AND short, wide ground paths to the bottom ground plane with small-profile, multilayer ceramic chip (MLCC) capacitors placed in close proximity to the IC.

6. Use the largest-size capacitor in the smallest body size you are assembling, at reasonable cost and with a voltage rating 2x the supply voltage. This is usually a 10 uF - 22 uF MLCC. Use only one value capacitor, but multiple capacitors in parallel, to reduce the mounting inductance and increase the capacitance.

3.9 Design for X

Along with DFC and DFP are all the other design goals aimed at reducing risk, cost, or schedule hits from re-spins in the board fabrication, assembly, or test, such as

- ✓ DFM: Design for manufacturing
- ✓ DFA: Design for assembly
- ✓ DFB: Design for bring-up, debug, troubleshoot
- ✓ DFT: Design for test
- ✓ DER: Design for reliability
- ✓ DFQ: Design for quality
- ✓ DFU: Design for the user

Collectively, all of these design goals are sometimes referred to as DFX: Design for excellence.

3.10 Practice Questions

1. What does connectivity mean?
2. Once connectivity is correct in a design, what is the most common general problem that can arise?
3. If the product works in a solderless breadboard, what does this mean about the interconnects?
4. What is the fundamental root cause of switching noise?
5. When does switching noise occur?
6. Why is it important to know the rise time of your signals?
7. What is the most important design guideline to follow to reduce ground bounce switching noise?
8. What is the most important design guideline to follow to reduce rail collapse switching noise?
9. What do you need to know to most efficiently fix a problem?
10. Why use the recommended minimum line width for signals?
11. What is a net class and why should you identify the net classes in the schematic?
12. What are the six most important design guidelines to follow to reduce noise?
13. Why do we say signal integrity lives in the white space and the wires of a schematic?

Chapter 4
Electrical Properties of Interconnects

When interconnects are not transparent, their electrical properties influence signal propagation and generate noise. To analyze and predict ahead of time the noise that might be present, we need to describe the real circuit components and the physical interconnects in terms of equivalent electrical circuit elements so we can estimate or simulate the expected noise.

4.1 Ideal vs Real Circuit Elements

Be careful when you use the term *ideal* to describe an interconnect. An ideal interconnect does not mean it is *transparent*. It does not mean that an ideal interconnect behaves like a wire element, as though it did not have any resistance or capacitance or inductance.

The term *ideal* refers to the circuit elements we use to describe the electrical properties. Each R, L, and C element in a schematic is ideal in that their properties are precisely defined in the circuit simulator. Usually, the simulator defines each ideal circuit element in terms of how the element treats the voltage across it, V, and the current through it, I.

The three common *ideal* circuit elements are defined by:

$$R = \frac{V}{I} \quad C = \frac{I}{\frac{dV}{dt}} \quad L = \frac{V}{\frac{dI}{dt}}$$

The value of the parameter: the resistance, R, the capacitance, C, and inductance, L, of each ideal circuit element, is each perfectly constant over frequency. These terms are the *parameters* that define each ideal circuit element, also referred to as their *figures of merit*.

A figure of merit is a powerful concept we will use over and over again in many engineering applications. It is one number that characterizes a circuit element or a behavior. It is usually a term that is intrinsic to the device and does not vary under external conditions.

If we know the figure of merit of a component, we know an important property of how it will behave.

For example, each parameter associated with an ideal circuit element is a figure of merit. The resistance of a resistor is one number that characterizes the behavior of that resistor. It is a figure of merit. The voltage of a battery is a figure of merit that describes the battery. A mutual inductance between two signal-return path loops is a figure of merit that describes the amount of noise that might be created.

Every ideal electrical circuit is composed of two parts: the various circuit elements contained in the circuit and how they are connected together, and the *parameter values* of each circuit element. We refer to how the various circuit elements are connected together as the *circuit topology*.

Our ability to accurately predict the voltage and current waveforms in a real circuit is limited by how accurate a model we can build using ideal circuit elements including their parameter values and the circuit topology.

We can only measure real components. We can only simulate or calculate with ideal circuit elements.

The general process we will follow to translate a real circuit we can measure into an ideal circuit we can calculate is to obey the guideline proposed by Einstein when he said, "Everything should be made as simple as possible, but not simpler."

Always start with the simplest model and grow in complexity only as needed.

It is truly remarkable that relatively complex real circuits can be accurately approximated with combinations of simple, ideal circuit elements. The process of taking a complex system and describing it with combinations of ideal circuit elements is called *strategic simplification*.

Every experienced engineer keeps these two worlds completely separate. There is the *real* world of *real* physical components and interconnects with some geometrical dimensions made from some combination of materials with material properties, and the *ideal* world composed ONLY of the Ideal circuit elements.

Train your engineer's mind's eye to see the equivalent circuit composed of ideal circuit elements when you look at the real components of an electronic product. When you look at the wires, see ideal resistors, inductors, and capacitors.

> *One of the most valuable skills of any experienced engineer is to be able to visualize the abstract world of ideal circuit elements whenever they look at real components.*

These two separate and distinct worlds are illustrated in **Figure 4.1**.

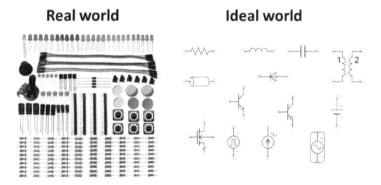

Figure 4.1 There are two separate and distinct world views, the real world and the ideal world. When you see the real components, think of the equivalent ideal circuit elements.

We can only build real prototypes and measure real voltages and currents using components from the real world. We can only build virtual prototypes and perform a calculation or simulation of predicted voltages or currents using ideal circuit elements.

A natural confusion arises when we use the terms resistor, capacitor, and inductor to describe both a *real*, physical component we assemble into a solderless breadboard or circuit board, and the *ideal* circuit elements used in a SPICE simulator. While we use the same names, these are not the same components. This is why using the preface *real* or *ideal* is so important to remove the ambiguity.

A real capacitor, for example, is a physical component with conductors in some shape and separation with some dielectric between them. The conductors don't even have to have a uniform shape, but can be a convoluted, 3D structure.

An ideal capacitor is very precisely defined. The only quality that defines any ideal capacitor element is one figure of merit, its capacitance. For an ideal capacitor, its capacitance is absolutely constant and never changes no matter what the rise time or frequency components of the voltages imposed on it. This is why the capacitance of an ideal capacitor is such a great figure of merit.

Likewise, a real resistor has some shape and material properties. An ideal resistor has just one parameter, its resistance. A real inductor is some conductor with some shape to it. An ideal inductor has just one parameter associated with it, its inductance. These distinctions between real components and ideal components are illustrated in **Figure 4.2**.

4.1 Ideal vs Real Circuit Elements

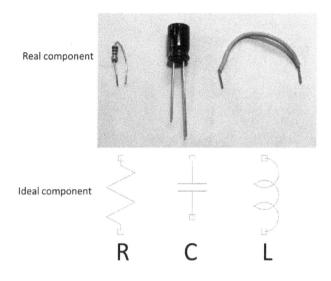

Figure 4.2 The distinction between real and ideal components.

The electrical properties of a real capacitor, such as its measured impedance, are approximated to first order, by an ideal capacitor. This is a perfectly fine ideal model to use in some cases. This is why an ideal capacitor model is so useful.

However, at higher frequency, the measured impedance of a real capacitor does not match the simulated, or predicted impedance, of an ideal capacitor. At this point, to achieve a better approximation, it is necessary to grow the complexity of the ideal model by adding more components. It is still an ideal model, it is just more complex.

To distinguish the two ideal models that differ in complexity but also in their accuracy in approximating the real component, we use the terms *first-order model* and *second-order model* or even *third-order model*.

For example, a first-order model of a real capacitor is a simple ideal C element. A second-order ideal circuit model for a real capacitor is a series RLC circuit. Each ideal element has an R, L, or C value constant with frequency. Yet their combined circuit impedance varies with frequency, and this is what matches the real measured impedance of a real capacitor.

When the simulated impedance of an RLC circuit matches the measured impedance of a real capacitor, the values of the R, L, and C elements become figures of merit to describe the real capacitor.

An example of the measured impedance of a real capacitor and the simulated impedance of an ideal C circuit and an ideal RLC circuit is shown in **Figure 4.3**.

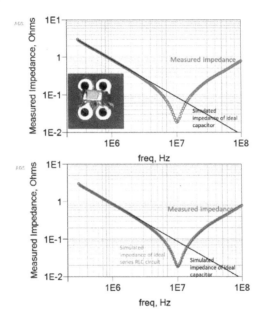

Figure 4.3 Top: Measured impedance of a real 0603 MLCC ceramic capacitor, shown in the inset, as the red circles. The black line is the simulated impedance of an ideal capacitor. Bottom: The same measured data with the simulated impedance of an ideal series circuit of ideal R, L, and C elements. The match is so good it is hard to see the simulated line.

The two different models to approximate the measured impedance of a real capacitor, the single ideal C circuit model, and the ideal RLC circuit model, are both ideal models. One is a first-order approximation, and the other is a second-order approximation of the real component.

EVERY model we use to approximate a real component is an ideal model. Just referring to an ideal model of a component is ambiguous.

4.1 Ideal vs Real Circuit Elements

We need to further clarify what circuit topology and parameter values we are using in the ideal model.

It is remarkable that combinations of these simple ideal circuit elements predict impedances, voltages, or currents that match the actual measured values in real circuits incredibly well. This is the value of ideal circuit models.

With accurate models and a circuit simulator such as SPICE, we can predict the voltage and current waveforms on any node of a circuit in the time or frequency domain that we would measure with a scope, for example.

Keep in mind that in a schematic, the real physical interconnects are always modeled as ideal wires. The schematic assumes the interconnects are transparent. This approximation may or may not predict the actual performance. This approximation may be too simple, in which case, the circuit's performance will depend on how we engineer the interconnects.

We may choose to approximate the properties of a *real* interconnect as an *ideal* wire element, which predicts there is no voltage drop across the interconnect no matter the current through it. An *ideal* wire circuit element is an *ideal* model of the interconnect, which also happens to be a transparent model.

We may choose to approximate the properties of a *real* interconnect using *ideal* R and L circuit elements, assuming the impact on circuit performance from an *ideal* capacitance is negligible. This is an *ideal* interconnect model.

We may choose to approximate the properties of a *real* interconnect in terms of just an *ideal* resistor element. This is an ideal interconnect model.

In order to assign a parameter value to each ideal circuit element, we need to know how the geometry and material properties of the real structures are translated into the parameter values. This process is covered in this chapter.

4.2 Equivalent Electrical Circuit Models

There are two families of ideal circuit elements we will use to approximate the behavior of interconnects depending on the rise time of the signals and whether the interconnects behave as electrically short or electrically long.

The circuit performance of a product is described in its schematic. In the schematic, interconnects are assumed to be electrically transparent. It is when the schematic is translated into the layout that the six types of noise come to life.

To analyze the impact the interconnects have on the electrical performance of the signals, we must first translate the physical geometry of traces with line widths, dielectric thickness, lengths, and the material properties, such as dielectric constant and conductor conductivity, into *equivalent electrical circuit elements*.

We can only estimate electrical performance using electrical circuit elements. Since the interconnects will only degrade the circuit performance and eat up performance margins, we sometimes call the electrical circuit elements that describe the interconnects as *parasitics,* in that they may suck the life out of a circuit.

The first step in analyzing the electrical performance is turning physical interconnects into circuit elements. We sometimes call this process *parasitic extraction.*

The propagating electromagnetic fields, which are really the signals, interact with the conductor and dielectric geometries of the interconnects. This behavior is completely described by Maxwell's equations, a collection of four equations that describe how electric and magnetic fields dynamically interact with the conductor and dielectric geometries or *boundary conditions.*

Maxwell's equations are the most accurate description of how signals propagate on interconnects and how noise is generated. If you want to become a master of signal integrity, you must eventually learn about and understand Maxwell's equations. See this article, for example: https://www.signalintegrityjournal.com/blogs/4-eric-

4.2 Equivalent Electrical Circuit Models

bogatin-signal-integrity-journal-technical-editor/post/1816-what-does-it-take-to-be-a-successful-si-pi-or-emc-engineer.

However, other than in the simplest situations, predicting the behavior of the electromagnetic fields from the initial conditions and the boundary conditions of the interconnects is very complicated.

In order to practically describe the signals and interconnects, we approximate the electromagnetic field description and describe signals as voltages and currents, and interconnects as equivalent electrical circuit elements. Inductors and resistors approximate how magnetic fields interact with interconnects, and capacitors approximate how electric fields interact with interconnects.

There are two general classes of circuit elements we use to describe interconnects: *lumped* circuit elements and *distributed* circuit elements.

The lumped circuit elements are the four circuit elements available in all circuit simulation tools, such as SPICE:

- ✓ An ideal resistor, R
- ✓ An ideal capacitor, C
- ✓ An ideal inductor, L
- ✓ And ideal mutual inductor, M

The simplest equivalent lumped circuit model for a signal-return path interconnect is shown in **Figure 4.4**.

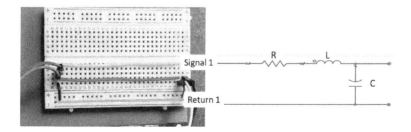

Figure 4.4 A physical interconnect and its equivalent lumped circuit electrical model.

The lumped circuit model is a good approximation for the actual measured behavior of interconnects when the interconnects are electrically short compared with the wavelength of the highest frequency component of the signal. In this regime, the electric and magnetic fields do not vary much across the length of the interconnect, and these electrical properties can be considered all lumped in one point. This criterion is shown in **Figure 4.5**.

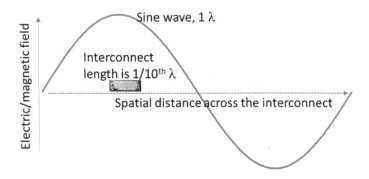

Figure 4.5 An example of the distribution of electric and magnetic field magnitude down a path compared to an interconnect that is 1/10th of a wavelength.

When we draw the sine wave representing the signal wave on the interconnect, it is a plot of the magnitude of the electric field at each point along the interconnect at one instant in time. If the interconnect is short compared to the wavelength, this means the electric field does not vary much across the interconnect. The electric field is approximated as constant across the entire interconnect at any one instant in time.

The electrical properties of the interconnect behave as though all of its properties are lumped into one point. This is the criterion for an electrically short interconnect.

This condition of electrically short, where the lumped circuit model applies, is:

$$\text{Len} < \frac{1}{10}\lambda \quad \text{and} \quad \lambda = \frac{v}{f} = \frac{v}{BW}$$

Where:

Len = the condition of electrically short interconnect length

λ = the wavelength of the highest sine wave frequency component of the signal

v = the speed of light in the dielectric of the interconnect

f = the highest frequency copoint of the signal

BW = the bandwidth of the signal— also the highest sine wave frequency component in the signal

A good estimate for the bandwidth (BW) of a signal, based on its 10%-90% rise time, RT, is (see for example, https://www.edn.com/rule-of-thumb-1-bandwidth-of-a-signal-from-its-rise-time/):

$$BW = \frac{0.35}{RT}$$

These combine to relate the longest interconnect length that is still electrically short and the 10%-90% rise time as:

$$\text{Len[in]} < \frac{1}{10}\frac{v}{BW} = \frac{1}{10}\frac{6}{0.35} RT[\text{n sec}] = 1.7 \times RT[\text{n sec}]$$

This assumes a speed of light in the dielectric of 6 in/nsec, which is a good estimate for most PCB laminate materials.

This suggests that for a 10 nsec rise time, the interconnects are electrically short, and a lumped circuit model is a good approximation for interconnects that are 17 inches or shorter. This is the case for many microcontroller signals on solderless breadboard interconnect wires.

When an interconnect is electrically short, it does not mean it is transparent, it just means it can be accurately described by a lumped circuit model.

When the interconnect lengths in inches are > 1.7 x rise time in nsec, the lumped circuit elements are not a very good approximation. At these longer lengths, compared to the wavelength, the interconnects behave electrically long. At these higher bandwidths or shorter rise times, we need to use distributed transmission line models.

Watch this video and I will walk you through comparing the measured impedance of a PCB trace with an ideal C and L model to show, up to what frequency it behaves electrically short.

4.3 Parasitic Extraction of R, L, and C Elements

The capacitance and inductance between a signal and return path for most interconnect geometries are difficult to calculate because their values are dominated by the precise shape of the fringe electric and magnetic fields.

While it is possible to use a 2D or 3D electromagnetic field solver to calculate the capacitance of any arbitrary structure, there are a few simple approximations for specific geometries. These have been implemented as online calculators on a number of websites. For example, this one created by Prof. Todd Hubing's students at Clemson University is simple to use: https://cecas.clemson.edu/cvel/emc/calculators/TL_Calculator/index.html

This tool calculates the resistance, capacitance, and loop inductance of the signal-return path conductors for four geometries. **Figure 4.6** shows an example of the geometry of two side-by-side rectangular conductors for a frequency below 1 MHz.

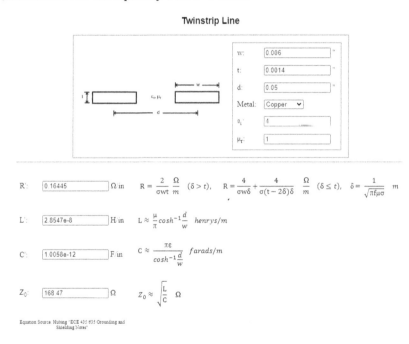

Figure 4.6 Example of the online calculator from Clemson University for twin strip lines.

In this example, for two strips, one is assumed to represent the signal and the other the return. The parasitic values are not affected by which is which. In this example, each trace is 6 mils wide in 1 oz copper, 1.4 mils thick, on 50 mil centers. They would have the following parasitic values:

R = 0.16 ohms/inch for both lines in series

L = 28 nH/inch loop inductance of the pair, or 14 nH/inch of total inductance of each line in the signal-return path pair

C = 1 pF/inch between the signal and return.

Calculating the loop inductance of a signal-return path is, in general, difficult. There are other approximations for the inductance of other signal-return geometries. For example, for a circular loop, the loop inductance is:

$$L_{loop} = 1.3uH/m \left(\frac{D}{2}\right)\left(\ln\left(\frac{8D}{d}\right) - 2\right) = 33nH/in \left(\frac{D}{2}\right)\left(\ln\left(\frac{8D}{d}\right) - 2\right)$$

This approximation has been implemented as an online calculator on a number of sites, such as this one:
https://www.eeweb.com/tools/loop-inductance/.

An example of using this calculator is shown in **Figure 4.7**.

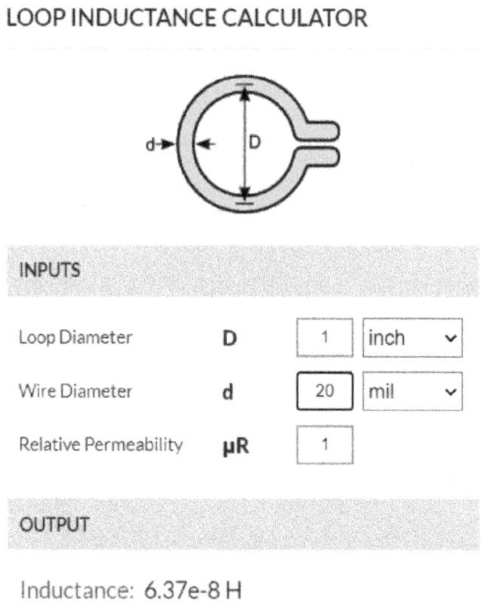

Figure 4.7 Using the online calculator on the EEweb site to estimate the loop inductance of a ring conductor.

In this example, a 1-inch diameter loop of 24-gauge wire has a loop inductance of 64 nH. The circumference is 3.14 inches. This is roughly a total inductance of 64 nH/3.14 inch = 20 nH/inch of circumference. This is the origin of the rule of thumb that the total inductance of a conductor is about 20 nH/inch of circumference when it is part of a loop.

This relationship highlights the important terms that influence the loop inductance of a circular loop. The larger the diameter of the loop, D, the larger the loop inductance. However, it does not scale with the area, D^2, but with $D \times \ln(D)$. This is clear when this approximation is plotted, as shown in **Figure 4.8**.

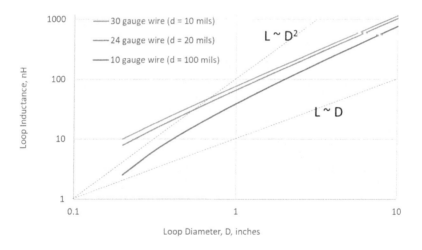

Figure 4.8 The approximation for loop inductance and loop diameter for three different wire diameters. The approximation is compared with estimates of the loop inductance scaling with the area or with the circumference. The actual variation is in between.

The wider the diameter of the wire, d, the smaller the loop inductance. But it is a soft dependence, in this geometry, scaling with ln(d). This plot also shows the loop inductance of three different loops made with 30 gauge (d = 0.01 in), 24 gauge (d = 0.02 in), and 10 gauge (d = 0.1 in) wires. A complete list of the wire diameters for different gauges can be found on Wikipedia, for example.

The wider the conductor cross section, the lower the loop inductance. This is an important property of loop inductance.

This specific online calculator has other structures built in, as shown in **Figure 4.9**.

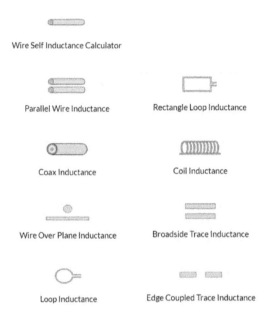

Figure 4.9 Examples of the types of geometries for which there are approximations implemented on the EEweb online calculator.

4.4 Describing Cross Talk

When interconnects are electrically short, the interconnect properties can be accurately approximated by the lumped circuit elements.

The cross talk between two electrically short signal-return path interconnects, driven by the electric and magnetic field coupling between them, can be approximated as capacitive, inductive, and resistive coupling.

4.4 Describing Cross Talk

The circuit element that describes inductive coupling or mutual inductance is a transformer element, shown in **Figure 4.10**.

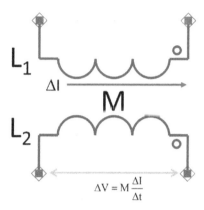

Figure 4.10 A transformer describes the coupling between two inductors.

It consists of two inductors with loop self-inductances, L_1 and L_2, with a loop mutual-inductance, M, between them. This is an excellent model when there are two signal-return path loops in proximity. The figure of merit that describes the coupling is either the mutual inductance, M, or the coupling coefficient, k:

$$k = \frac{M}{\sqrt{L_1 L_2}}$$

This defines the voltage generated across the second inductor, V_2, when there is a changing current in the first inductor, I_1:

$$V_2 = M \frac{dI_1}{dt}$$

If the mutual inductance is 5 nH, the changing current is 20 mA in 5 nsec, the voltage induced across the second loop during the rising edge is

$$V_2 = M \frac{dI_1}{dt} = 5\text{nH} \frac{20\text{mA}}{5\text{n sec}} = 20\text{mV}$$

The size of the loops and the proximity of the signal-return loops affects the mutual inductance. The larger the loop areas, the larger the mutual inductance between the two loops.

In addition, the closer the two loops, the larger the mutual inductance. In the extreme case, when the return paths of the two loops are shared, the mutual inductance can be as large as half the loop inductance of either loop.

When the coupling is dominated by a shared return path, the mutual inductance is the shared total inductance of the return path. All the magnetic field lines around the aggressor's return path are also around the victim's return path and are shared. This geometry will always give a larger mutual inductance than when the return paths are separate conductors. These two circuit models are shown in **Figure 4.11**.

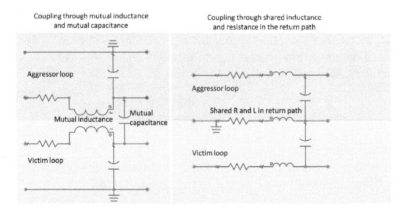

Figure 4.11 Two different models for cross talk. Left: when the signal-return loops are separate. Right: when the return paths are shared.

An example of a data line and switching noise on a quiet line is shown in **Figure 4.12**.

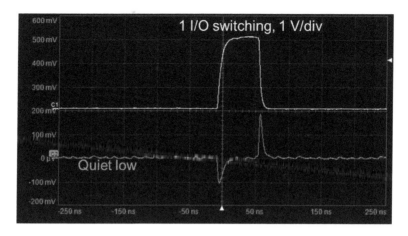

Figure 4.12 Measured output digital signal from a 328 uC pin and the switching noise on an adjacent digital pin.

Watch this video and I will demonstrate mutual inductive cross talk between two loops.

4.5 Estimating Mutual Inductance

In the absence of an electromagnetic field solver, it is difficult to estimate the mutual inductance between two loops. It depends very much on all of their geometric terms. Using a 2D field solver, we can explore the mutual inductance of two conductors to estimate the value of the coupling coefficient, k.

In this example, two wide conductors, one on top of the other, are adjacent to two other conductors with similar geometry. These are considered broadside coupled. Their configuration is shown in **Figure 4.13**.

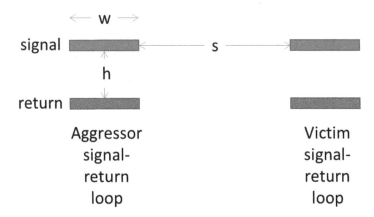

Figure 4.13 Setup to estimate the mutual inductance between the two signal-return loops using a 2D field solver.

In this example, a 2D field solver integrated in Keysight's ADS simulator was used to calculate the loop self-inductance of each loop and the loop mutual inductance between them, as the spacing was changed for three different heights, h.

The calculated coupling coefficient, k, for the case of 10 mil line widths, 1 mil thick, as the spacing between the two pairs of traces are moved apart, is shown in **Figure 4.14** for a height of 5 mils, 10 mil, and 15 mils.

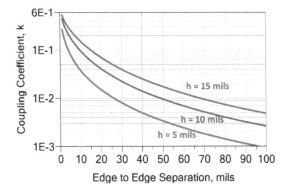

Figure 4.14 Increasing the edge-to-edge spacing of the coupled pairs.

4.5 Estimating Mutual Inductance

For the special case of 10 mil wide traces, when the separation between the signal and return trace, h, is also 10 mils, the coupling coefficient is 10% when the edge-to-edge spacing is 10 mils.

When the spacing is h = 5 mils and the edge-to-edge spacing is 5 mils, the coupling coefficient is 10%.

When the spacing is h = 15 mils and the edge-to-edge spacing is 15 mils, the coupling coefficient is 10%.

This suggests that when the spacing between two loops is comparable to the distance between the signal and return path, the coupling coefficient is 10% of the self-inductance of either one. This is a rough estimate of a starting place for the coupling coefficient for two loops.

And when the two loops are moved farther apart, the mutual inductance drops off much faster.

This calculation illustrates the two important principles of how mutual inductance varies with geometry:

- ✓ As the separation between the two loops increases, the loop mutual inductance quickly decreases in this geometry.
- ✓ For a fixed spacing between the loops, as the signal and return paths get closer together, the loop self-inductance of either pair will decrease, the mutual inductance will decrease, and the coupling coefficient will decrease.

To reduce the loop mutual inductance between two separate loops, the design guideline is to bring the signal and return of each loop closer together, which reduces the loop self-inductance of each loop, and to pull the two loops farther apart.

As a rough starting place, the coupling coefficient is 0.1 when the spacing between the loops is comparable to the spacing between the signal and return in either loop.

When there is a shared return path, the coupling coefficient can be as large as 50% if half of the loop is part of the shared path. This is much larger than the typical mutual inductance between two loops and is why shared return paths can dramatically increase mutual inductance.

4.6 Training Your Engineer's Mind's Eye

When you look at a solderless breadboard or a circuit board, you see the physical interconnects. Maybe you see the individual signal and return paths.

In most interconnects with individual wires connecting terminals, it is the inductance of the interconnects — the loop self-inductance of the signal-return path pairs and the loop mutual inductance between adjacent signal-return path pairs — that dominates the electrical performance of the interconnects.

If you care about reducing the noise generated by interconnects, whenever you see a real wire, see with your mind's eye an inductor of about 20 nH per inch of the wire. This will increase your awareness of the importance of the interconnects influencing noise generation.

Instead of seeing some wires, you should learn to train your engineer's eye to see the equivalent circuit elements connecting the terminals. You should learn to see the small inductors connecting terminals and the small mutual inductance between adjacent loops.

As a simple, first-order estimate, imagine each inch of wire being replaced with a 20 nH inductor.

Between loops, if the return paths are separate, replace the two loops with two inductors and use a value of the mutual inductance between the loops of about 10% the loop inductance of either loop.

If there is a shared return path, then use 20 nH of mutual inductance per inch of shared path as the mutual inductance.

For example, when there are discrete wires connecting a resistor into a circuit, as shown in **Figure 4.15**, replace the real resistor and its wires in your mind's eye with an ideal resistor and an ideal inductor having a loop inductance of 20 nH per inch of circumference of the path.

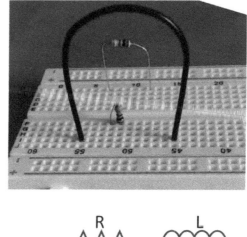

Physical world

What an engineer sees

Figure 4.15 Examples of the physical structures and what an engineer sees in their mind's eye. Each structure has a different value of R and L but a similar equivalent circuit model.

This process of imagining the inductance will train you to see the loop inductance in the power distribution path and the mutual inductances in the signal paths. You will become sensitive to how important optimizing the geometry is when designing interconnects to reduce their noise.

4.7 Electrically Long Interconnects

The distributed circuit elements, also available to some extent in all circuit simulators, are the transmission line or T elements.

All interconnects can always be approximated by transmission line elements. These models match measured performance of real interconnects from DC to very high bandwidth. Single-ended transmission line models can accurately describe all forms of self-

110 *Practical Guide to Prototype Breadboard and PCB Design*

aggression noise and coupled transmission line models can describe all forms of mutual-aggression noise or cross talk.

For example, in a circuit with an output pin and an input pin, when the rise time is short enough and the interconnect long enough, with the source impedance of the transmitter low enough and the receiver impedance high enough, the signal at the receiver will show ringing noise. (Note all the special conditions, which are not uncommon conditions.)

This is fundamentally due to the distributed properties of the interconnects, accurately predicted by the transmission line properties of the interconnect. An example of this ringing noise measured with a scope and the same physical system simulated with a SPICE simulator using a transmission line model is shown in **Figure 4.16**.

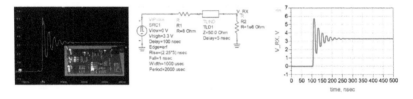

Figure 4.16 An example of the measured signal at a receiver pin from an output pin in the circuit board shown in the insert. Left: the measured voltage on the receiver pin. Middle: the transmission line circuit in Keysight's ADS. Right: the simulated voltage on the receiver pin. The simulation waveform matches the measured waveform very closely.

The ringing noise is from the signal interacting with its own interconnect and this interconnect adding noise to the signal. It is a form of *self-aggression* noise. Fundamentally, it is due to reflections of the signal bouncing off the ends of the interconnect, diminishing with each bounce. This is a fundamental property of signals interacting with transmission lines and impedances.

The way to reduce this sort of problem is to design all the interconnects as controlled impedance transmission lines, use a termination strategy at either of the two ends, and minimize the branch lengths in the routing.

For additional information on the properties of transmission lines, see the book Bogatin's Practical Transmission Line Design and Characterization for Signal Integrity Applications.

While the distributed transmission line models are the most accurate interconnect models over the widest bandwidth, they are sometimes complicated to use correctly. Transmission lines are an advanced topic. If you expect to advance in circuit design and deal with the highest-performance systems, understanding transmission lines must be in your future.

If interconnects are electrically short, their electrical properties can be approximated by the lumped circuit elements. These circuit models dramatically simplify the electrical analysis and are the focus of this book.

The condition for an interconnect to be considered electrically long, where transmission line properties dominate interconnect performance, is that the electrical properties of the interconnects can only be accurately approximated by transmission line models. Lumped element models are not an accurate description of the interconnects when the interconnects are electrically long.

There are a number of ways of estimating this transition, but they should all be considered just estimates or rules of thumb and not a hard limit. This transition between electrically short and long is just about which is the better equivalent circuit model to use.

While distributed transmission line models are an accurate representation of all interconnects from DC to some high bandwidth, they carry a higher cost in complexity to understand and to use and not all simulators have accurate transmission line models. This is especially the case when describing cross talk between an aggressor and victim signal-return path loop.

4.8 Electrically Short and Electrically Long

Being electrically short does not mean the interconnects are transparent. Being electrically long does not mean they must be terminated for acceptable performance. This condition just refers to the appropriate equivalent circuit model to use.

Once we determine if an interconnect is electrically long, we can use the model of the interconnect and a simulation to evaluate whether a termination strategy is required based on the combination of driver models, routing topology, and reflection noise generated compared to what is acceptable.

The condition for electrically short depends on both the rise time of the signals and the length of the interconnects. In the last section, we estimated this connection between physical length of the interconnect and the rise time of the signal based on when the length was shorter than 1/10th a wavelength. This condition was:

$$\text{Len}_{\text{long}}[\text{in}] > 1.7 \times \text{RT}[\text{n sec}]$$

Another way of evaluating this threshold of when an interconnect is electrically long and transmission line analysis is important is by looking at the impact from ringing noise due to reflections based on the transmission line properties of the interconnects.

As the signal rise times decrease or lengths increase, the ringing noise from the reflection properties of the transmission line behavior of the electrical long interconnects on the interconnects due to the distributed transmission line effects become more pronounced and a transmission line description becomes essential. This is one of the conditions that can be used to describe the threshold for an interconnect to be electrically long.

For example, **Figure 4.17** shows the measured signal at an oscilloscope input as a receiver from a fast, low impedance source as the length of the interconnect increases. Above a threshold, as the interconnect length increases, the ringing noise from transmission line effects increases.

4.8 Electrically Short and Electrically Long 113

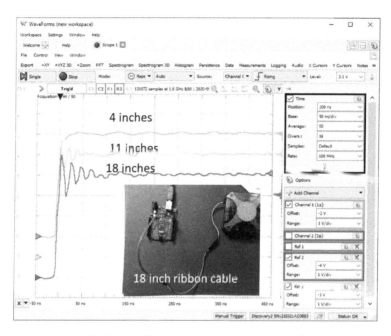

Figure 4.17 An example of the reflection noise in an interconnect from the signal source on a microcontroller board to the input to a scope for three interconnect lengths. The longer the interconnect, the more the reflection noise.

Reflections on interconnects from impedance changes will always happen. However, when an interconnect is short enough and its time delay is short compared to the rise time of the signal, the resulting reflection noise may be low enough to not be a concern.

The magnitude of the reflection noise depends not just on the rise time of the signal and the time delay of the interconnect. It also depends on the output impedance of the driver. As the starting place, we make the assumption that the output impedance is very low compared to 50 ohms.

The signature of reflection noise looks like a damped sine wave. The period of the sine wave voltage noise is related to the time delay, TD, of the signal to propagate from the transmitter pin to the receiver pin. The signal reflects from the high impedance RX, heads to the TX, reflects and changes sign from the low impedance of the TX, hits the RX again, then heads back to the TX, reflects and changes sign and comes back to the RX again with a positive peak. The period of ringing is four one-way time delays:

$$\text{Period} = 4 \times \text{TD}$$

When the rise time is long compared to the ringing period, the reflections will smear out during the rising or falling edge of the signal. In this case, the reflections will always happen, but they may not be visible and would play no role affecting signal quality at the receiver. In this case, the transmission line properties of the interconnect can be approximated by the lumped circuit elements.

The amount of ringing depends on many details. It can overwhelm any signal, or it can be insignificant. An example of a typical case and the resulting signal at the receiver for three different rise times is shown in **Figure 4.18**.

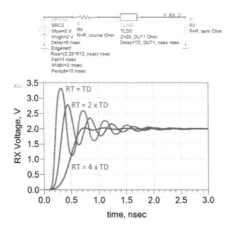

Figure 4.18 Top: example of a simple transmission line circuit with a signal source with a 10 ohm source impedance, driving a transmission line with a high impedance receiver. Bottom: the simulated voltage at the receiver as the rise time of the signal increases compared to the time delay of the transmission line. Simulated with Keysight's ADS.

When the rise time is shorter than 4 x TD, the reflections can be significant, and the transmission line properties of the interconnects are important. This is the condition for an electrically long interconnect when transmission properties are important.

4.8 Electrically Short and Electrically Long

But, if the rise time is longer than 4 x TD, the transmission line properties are not important. This is the condition for an electrically short interconnect when transmission line effects are not important.

Generally, in a printed circuit board, the typical laminate materials have a dielectric constant of about 4 and the speed of a signal is about 6 inches/nsec. For an interconnect of length, Len, the time delay, TD, is:

$$TD[n\sec] = \frac{Len[in]}{v} = \frac{Len[in]}{6 \text{ in}/n\sec}$$

As a rough rule of thumb, an interconnect is electrically short when the transmission line properties of the interconnect are not significant when:

$$RiseTime[n\sec] > 4 \times TD = 4 \times \frac{Len[in]}{6 \text{ in}/n\sec} = 0.7 \times Len[in]$$

or

$$Len[in] < 1.4 \times RT[n\sec]$$

For example, if the rise time is 3 nsec, an electrically short interconnect, when transmission line effects are not important and a lumped circuit model is adequate to describe the interconnect, is for an interconnect length shorter than 1.4 x 3 nsec = 4.2 inches.

The electrical properties of interconnects shorter than 4.2 inches can be accurately described with lumped circuit elements.

This is remarkably close to the estimate of an electrically short interconnect based on when the interconnect length is 1/10th of a wavelength. This suggests that a good rule of thumb for when interconnects are electrically long is roughly:

$$Len_{long}[in] > 1.5 \times RT[n\sec]$$

If you worry whether this relationship should have a 1.4 or 1.7 or even a 1 as the scaling term, then don't use this criterion. You should assume the interconnect is electrically long and use a transmission line model to evaluate the impact of the interconnect on electrical performance.

This relationship between the interconnect length and the rise time for when transmission line properties play a role is shown in **Figure 4.19**.

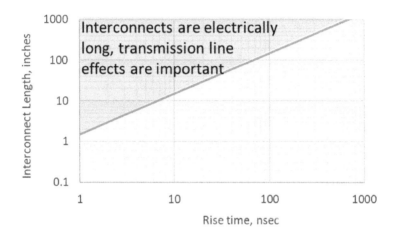

Figure 4.19 The boundary for when interconnects are electrically long, based on the criteria of Len[in] > 1.5 x RT[nsec].

For example, when the rise time of the signal is 10 nsec, interconnects shorter than about 15 inches are electrically short. This does not mean that if rise times are 5 nsec and the interconnect lengths are longer than 15 inches, the interconnect will cause a problem. It just means you will get a better estimate of the reflection noise using transmission line models of the interconnect.

Whether ringing self-aggression noise is significant depends on:

- ✓ The rise time of the TX
- ✓ The output source impedance of the TX
- ✓ The length of the interconnects
- ✓ The characteristic impedance and uniformity of the interconnects

4.8 Electrically Short and Electrically Long

- ✓ The routing topology of the interconnects
- ✓ The impedance of the RX
- ✓ The noise margin: how much noise is too much

When interconnects are electrically long, transmission line analysis should be performed to determine whether it is necessary to implement some of the design guidelines to reduce reflection noise to an acceptable level, such as designing the interconnects as uniform, controlled impedance transmission lines, implementing a termination strategy, reducing the branch lengths in the routing topology, or even limiting the routing topology as point to point or daisy chain.

The first step in any design is to determine whether your interconnects are transparent or not and then if they are not, can they be approximated with lumped circuit elements or distributed transmission line elements. Which regime a product is in depends on the rise time of the signals and the interconnect lengths. These relationships are shown in **Figure 4.20**.

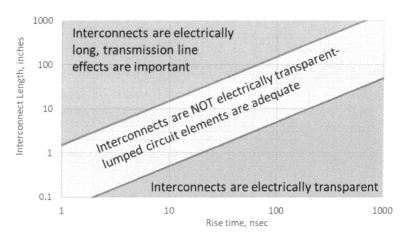

Figure 4.20 Mapping design space into the three regions of transparent, where lumped elements can be used and when transmission line elements are needed.

Another way of identifying the range of electrically long or electrically short interconnects is to consider the behavior of the signal propagating on the transmission line.

When the signal is originally launched on the interconnect, the voltage between the signal and return path increases with the rising edge. As the voltage turns on and increases, the signal launched on the transmission line will begin to propagate down the line at the speed of light in the materials, v.

Once the rising edge has reached its peak, the signal stops changing, and the edge continues to propagate down the transmission line. This rising edge, as it propagates, has a spatial extent on the transmission line, as illustrated in **Figure 4.21**.

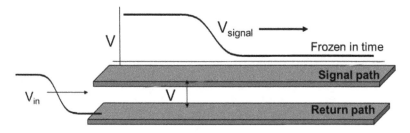

Figure 4.21 An illustration of the spatial extent of the rising edge of the voltage wave on a signal-return path interconnect.

In the time of the rising edge, the beginning of the edge has moved down the interconnect a distance of RT x v. This is the spatial extent of the rising edge.

When the length of the interconnect is longer than the spatial extent of the rising edge, the interconnect is considered to be electrically long: the interconnect is long compared to the spatial extent of the rising edge of the signal.

This condition for an electrically long interconnect translates to:

$$\text{Len} > \text{RT} \times v = \text{RT} \times \frac{c}{\sqrt{Dk}} \sim \text{RT} \times 6\text{in}/\text{nsec}$$

This condition for electrically long means Len[in] > 6 x RT[nsec].

This is a much more liberal criteria than introduced earlier, that electrically long means Len[in] > 1.5 x RT[nsec]. This suggests that while this criterion is a simple way of visualizing the behavior of an electrically long interconnect, transmission line effects will potentially play a role at lengths that are shorter than this criteria.

4.9 Practice Questions

1. What is special about an ideal circuit element?

2. What is special about a real circuit element?

3. How does the inductance of an ideal inductor vary with frequency?

4. What is the difference between a first-order ideal model and a second-order ideal model?

5. What can you do with a real circuit element you can't do with an ideal circuit element?

6. Why do we try to model real circuits with ideal circuit elements?

7. Why is it ambiguous if we describe an interconnect as an "ideal" interconnect? What does this refer to?

8. What are the four common lumped circuit elements?

9. What is a distributed circuit element?

10. What does it mean to have an electrically short interconnect?

11. What does it mean to have an electrically long interconnect?

12. What is an estimate of the inductance of a length of wire 2 inches long that is part of a loop?

13. How is the period of the ringing in an unterminated transmission line related to its time delay?

14. The signal rise time is 1 nsec. Up to what length interconnect could be modeled as an inductor or capacitor?

15. If an interconnect is electrically short and the RX is low impedance, like an LED, what is a good model for the interconnect?

16. If an interconnect is electrically short and the RX is a high impedance, like the input to a MOSFET, what is a good model for the interconnect?

Chapter 5
Trace Width Considerations: Max Current

One of the very first questions to ask when beginning a circuit board layout is, *what line width should be used for signals and for power or ground routing?* What is the *best design practice?*

If the design requires a controlled impedance circuit board, the target characteristic impedance, dielectric constant of the laminate, and the dielectric thicknesses of each layer determine the required line width. For example, for a microstrip surface trace, if the target impedance is 50 ohms and the Dk is 4, with a dielectric thickness of 10 mils, the line width should be about twice the dielectric thickness, or about 20 mils.

But, if interconnects are electrically short, their characteristic impedance is not an important design constraint. The characteristic impedance of a trace is not an important driving force controlling the line width.

In this case, the design guideline for the *default* signal net class, unless you have a strong compelling reason otherwise, should always be to use the narrowest line possible to provide the highest interconnect density and easiest routing.

There are three considerations that limit how narrow the line width can be. In order of the most important criteria, they are:

- ✓ No narrower than the selected fab vendor can manufacture at no cost premium. This is typically 6 mils for all low-cost fab shops.

- ✓ Use a wide enough line to carry the maximum DC steady state current without heating up above an acceptable temperature threshold.

✓ Use a wide enough line width to provide a low-enough series resistance acceptable for the application.

Each of these criteria are evaluated below.

5.1 Best design practices

We will use the concept of *best design practices* over and over again. These are the generally accepted guidelines we will follow as long as there is no strong compelling reason otherwise.

Don't just follow rules someone gives you. Understand why we have specific *best design practices* so that you can determine when they apply, and when you might have a strong compelling reason not to follow them.

> *If you are not going to follow the best design practices, you should be aware of the strong compelling reason why not.*

5.2 Minimum Fabrication Trace Width

Generally, most board shops will offer a capabilities page that shows what line widths and spacing between traces, referred to as the minimum track/space feature, they can fabricate at no additional cost.

Fabricating a narrower line/space will add cost either due to using a different photoresist or etching process, more expensive process monitoring, or assuming some production yield hit in the boards.

Before you lay out your circuit board, you should select a fab shop and make note of their fabrication capabilities. If you do not have a specific fab shop in mind, a good starting value for the minimum line width all fab shops, even the low-cost internet fab shops, can do at no extra cost is 6 mil track and space.

For example, the options for www.PCBway.com are shown in **Figure 5.1**. The price for 5 boards, 100 mm x 100 mm with the default values and 6 mil line and space, is $5. If the 5-mil line and space is select, the price increases to $40.

Figure 5.1 Example of the features PCBway is capable of fabricating at no extra charge.

Unless you have a strong compelling reason otherwise, do not use a line width or space narrower than 6 mils. If you do, your board may cost more, and you may not be getting any additional value for the higher price.

Watch this video and I will walk you through evaluating the fabrication capabilities of a few fab shops.

5.3 Copper Thickness as Ounces of Copper

The line width of copper is measured in mils or mm. The minimum line width commonly used is 6 mils or about 0.15 mm. The thickness of copper can also be measured in units of mils or mm. However, it is more commonly measured in units of ounces or oz.

In the early days of printed circuit board fabrication, the plating thickness of copper on a board was measured by weighing the board. The weight of plated copper per square foot of board area on the board is related to the copper thickness based on the density of copper:

$$\text{Weight} = \text{density} \times A \times t \quad \text{or} \quad t = \frac{\text{Weight}}{\text{density} \times A}$$

The units are a little unusual. The weight will be in oz. The area, A, will be in ft² and we want the thickness in microns. This means the units for density should be in units of oz/ft²-u. We can do the units conversion for density very easily in steps:

$$\text{density} = 8.9 \text{gm/cm}^3 \times 0.035 \text{ oz/gm} \times 30.5^2 \text{cm}^2/\text{ft}^2 \times 10^{-4} \text{cm/u} = 0.029 \text{oz/ft}^2 - u$$

This makes the thickness for 1 oz copper to be,

$$t = \frac{\text{Weight}}{\text{density} \times A} = \frac{1 \text{oz}}{0.029 \text{ oz/ft}^2 - u \times 1 \text{ft}^2} = 34 \text{ u} = 1.3 \text{ mils}$$

This thickness will vary depending on the density of copper and is often reported as between 1.2 mils to 1.4 mils in thickness.

If it is important to know the precise thickness of the copper, the traces should be cross sectioned.

For example, a plating weight of 1 oz of copper per square foot of board area corresponds to about 34 um or approximately 1.3 mils thick plating. This means ½ oz copper has a thickness of 17 um or about 0.7 mils.

A standard thickness for copper on outer layers is 1 oz. In some processes, all signal traces on outer layers are plated up an additional 1 mil when the through-hole vias are plated. This means the signal traces are actually 2.3 mils thick.

In other processes, the board surface is coated with a plating resist, except where the through-hole vias are. The via barrels are the only features plated with extra copper, leaving the traces as the original foil thickness of 1 oz.

Inner layers in a four or more layer board use copper foil typically at either ½ oz or 1 oz thick. The inner layers are not plated up when through-hole vias are exclusively used for the board.

5.3 Copper Thickness as Ounces of Copper

Selecting a thicker copper for an inner or outer layer will generally cost more. For example, with PCBway, moving from 1 oz copper to 2 oz copper on an outer layer, the price of 5 boards increases from $5 to $49. Unless you have a strong compelling reason otherwise, stick with the default values of 1 oz copper.

Generally, there are three reasons thicker copper might be needed in a design.

Obviously, thicker copper will have lower electrical resistance than thinner copper. Those applications where very low resistance is needed, generally either high current or when very low DC voltage drops are required, might benefit from thicker copper. However, a similar benefit might be gained by just using a wider trace at no added cost.

In addition to lower electrical resistance, the other electrical benefit with thicker copper is higher current carrying capability before a significant temperature rise. Current loads to a specific device of more than 40 A DC current might benefit from copper foil thicker than 1 oz copper. Again, the same benefit might be possible with a wider trace with no added cost.

In many applications, the copper planes in a board provide not just electrical conductivity but also thermal conductivity. The planes, and even wider traces, can be used as heat spreader elements.

When you have a high-power consumption device, like a transistor or MOSFET operating in the linear regime, dissipating 1 watt or more, it might be useful to engineer a wide surface trace under the component or a solid plane under the component that is well thermally coupled to the component to suck the heat away and spread it to a larger area from which normal air convection can remove the heat from your board. Thicker copper might help remove more of the heat in these applications.

Whenever making the decision to use thicker copper compared with the standard 1 oz copper, it is important to compare the value returned at the higher cost. If there is no added value with the more expensive feature, why would you want to pay more and get nothing in return? This is a waste of money.

5.4 Maximum Current Handling of a Trace

A 6-mil wide trace is really narrow. While its width is as narrow as a human hair, its thickness is much less than a human hair. Surely, such a tiny cross section of copper can't carry much current before it gets too hot or even melts.

Right now, grab a piece of paper and write down your guess of how much DC current you think you can send through a 6-mil wide, 1 oz copper trace before it gets too hot. The real answer will surprise you.

When we put some current in a wire, it consumes power due to joule heating and heats up. Power *consumption* is the conversion of electrical energy into thermal energy by a component or device. Ultimately, power is only consumed by resistive elements, having some resistance, R. The power consumption in an ideal resistor is:

$$P_{consumption} = I^2 R$$

In this example, the resistive element that consumes power is the resistance of the trace. The higher the consumed power, the more the trace will heat up and its temperature will rise.

The higher the temperature, the more heat flows from the trace into the environment to cool the trace. While power *consumption* is the heat power created by the device, power *dissipation* is the heat flow leaving the resistor into the ambient environment that cools the device.

In the simplest model, the power dissipation, the heat flow from a hot trace on a board to the ambient environment, either to the air or to the rest of the circuit board, is proportional to the temperature difference between the trace and the ambient, and inversely to the thermal resistance of the interface:

$$P_{dissipation} = \frac{\Delta T}{\theta}$$

The thermal resistance, Θ, describes the temperature rise for a given power flow in units of degC/watt. The higher the thermal resistance,

5.4 Maximum Current Handling of a Trace

the larger the temperature difference required to pump the same amount of heat power through the interface.

As constant power is consumed by the wire, it heats up. As it heats up, the temperature difference between it and the ambient increases and more power flows from the wire to the ambient. The power dissipation into the environment increases. Eventually a temperature difference is reached where the power consumed is equal to the power dissipated. This final, steady state temperature difference is:

$$P_{dissipated} = P_{consumed}$$

$$\frac{\Delta T}{\theta} = I^2 R$$

$$\Delta T = I^2 R \theta$$

The equilibrium temperature difference between the trace and the ambient, the temperature of the trace above ambient, depends on the thermal resistance of the trace environment. For a given trace geometry that defines the resistance and the power consumption for a specific current, we want to limit the current through the trace so that its equilibrium temperature rise above ambient is less than an acceptable value, typically between 10 degC rise for a very conservative value, up to 60 degC for a more extreme value.

A reasonable temperature rise above ambient that still poses no danger is 40 degC temperature rise.

The maximum current before a 40 degC temperature rise is strongly dependent on the features of the circuit board. The thermal environment of a trace on a board varies considerably from board to board and even trace to trace. It's really hard to calculate the final temperature unless you know all the details of the thermal environment and the power consumption of other nearby devices and traces.

But *sometimes an OKAY answer NOW! is more important than a good answer late.*

The IPC, an industry organization that manages a lot of PCB technology specs, sponsored a number of studies to build test boards and measure the maximum current you can put through a trace of a specific copper thickness and a width on an outer layer or inner layer of a board before exceeding a specific temperature rise.

They developed a few empirical equations to describe this max current. Their original specification, IPC 2221, was based on empirical measurements made in the 1960s. These experiments were revised and a new spec, the IPC 2152, was released in 2010 with more recent measurements.

The IPC recommended specification to follow for the maximum current capacity for circuit boards is the IPC 2152 spec. When using online calculators to estimate the current capacitance of PCB traces based on an IPC, be sure to check which spec is being used.

However, keep in mind that both specs are based on empirical measurements with many assumptions about the board stack-up and other features near the traces. These estimates should be used as rough guidelines, not hard-and-fast rigid rules to follow.

In addition to online tools that implement the IPC 2152 spec, a very useful calculator that runs under MS Windows can be downloaded from Saturn PCB (http://saturnpcb.com/pcb_toolkit/). This calculator uses the approximations in the IPC 2152 specification for its estimates.

An example of using this tool to estimate the maximum current through a 1 oz thick, 6 mil wide trace for a 40 deg temperature rise, is shown in **Figure 5.2**. The estimate is 1 A.

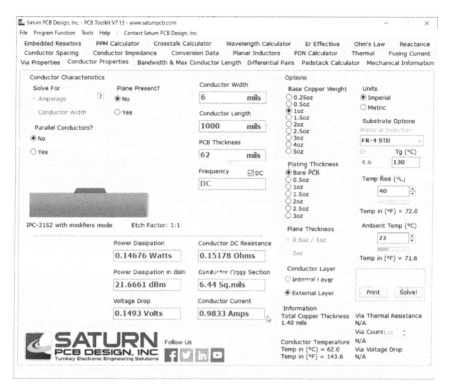

Figure 5.2 The result of estimating the max current for a 6 mil wide trace. This is only a rough approximation.

The way you use this calculator is first to input all the fixed terms, such as the base thickness before outer layer plating, the plating thickness, the PCB substrate thickness, the acceptable temperature rise, and any other terms. Then after clicking the Solve! button, the resulting maximum conductor current based on the IPC 2152 spec is calculated and displayed.

In this example, we found the maximum current we could send through the 6 mil wide, 1 oz copper trace before the temperature rise was 40 degC above ambient was about 1 A. This is only a very rough estimate, based on many assumptions about the geometry and thermal properties of the circuit board. It does not mean that if the current goes to 1.2 A, the temperature will rise above 40 deg C and the board will fail.

Watch this video and I will walk you through using the Saturn PCB tool to estimate the maximum current handling of a trace.

The estimates from the IPC 2152 spec are conservative estimates. As a simple experiment, a board was built with a variety of uniform lines with different line widths. A constant current was driven into these lines and the temperature was measured by touch. **Figure 5.3** shows a photograph of this test board. The top layer had 1 oz copper with trace widths of 6 mils, 8 mils, 10 mils, 20 mils, and 100 mils.

Figure 5.3 Example of the test board with different line widths, each 1 inch long fabricated with 1 oz copper. Note the connectivity pattern.

The bottom layer did not have a plane, but had a similar set of traces.

A 1 A current was sent through the 6 mil wide trace. Even after 2 minutes, the trace was not noticeably warm. The IPC 2152 spec suggested the temperature should have been 40 deg above ambient. The thermal environment of this trace in this specific board had much lower thermal resistance to the ambient than that of the traces in the IPC 2152 examples.

5.4 Maximum Current Handling of a Trace

In fact, the current was raised to 2 A before the trace was noticeably warm to the touch. At 3 A, the 6 mil trace was hot. At 4 A, the trace quickly heated red hot and fused open.

> *The estimate of 1 A as the maximum current to put through a 6 mil, 1 oz copper trace on an outer layer is a very conservative estimate and can be used as a safe design guideline. A 6 mil wide trace is perfectly suitable to carry up to 1 A of DC current.*

Using the same analysis with the Saturn PCB tool, a 20 mil wide trace is capable of handling 3 A of current before its temperature rise is above 40 degC above ambient. On this same test board, 3 A was sent through the 20 mil wide trace and again, there was no measurable temperature rise to the touch. It wasn't until 10 A was run through the trace that it heated up quickly and fused.

The estimates from the IPC 2152 spec and the simple empirical measurements on the test board confirm the following safe limitations:

A 6 mil wide trace can handle 1 A of DC current.

A 20 mil wide trace can handle 3 A of DC current.

This suggests a robust design guideline:

> *Route all signal traces, which typically carry less than 1 A of DC current as 6 mil wide lines, unless you have a strong compelling reason otherwise.*

> *Route all power traces, which might carry less than 3 A of DC current with 20 mil wide traces, unless you have a strong compelling reason otherwise.*

Using these guidelines means you will get the highest interconnect density without paying a premium price and you will be able to distinguish at a glance which traces on your board are carrying power distribution and which ones are carrying signals.

Even if your power traces will only carry a maximum of 100 mA, it is a good idea to route them as 20 mil wide traces. This will help identify them at a glance and aid in debugging a board.

5.5 Maximum Current Through a Via

The same analysis for the maximum current through a trace can be applied to find the maximum recommended current that can be sent through a via.

The Saturn PCB tool has a tab to calculate the IPC 2152 specification recommendations for a via. The conditions can be estimated for a typical drilled through-hole via, 13 mils in diameter. The conditions are shown in **Figure 5.4**.

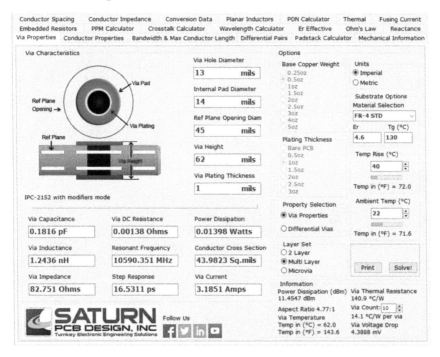

Figure 5.4 The screen shot from the Saturn PCB calculator showing a maximum current through a via for a less than 40 degC rise above ambient of 3.2 A.

The IPC 2152 recommendation for the maximum current that should be used in a 13 mil diameter through-hole via is less than 3 A. This is the same estimate as for a 20 mil wide trace. Of course, it depends on the local thermal environment of the via, but it is safe to assume the maximum current handling for a 13 mil via is 3 A. If more current is required in an application, such as when transporting a 10 A current, assume 3 A per via, which would require 4 vias in parallel between any power trace transitioning from one layer to another, which would carry the full 10 A.

5.6 Thermal Runaway with Constant Current

There are generally two types of power sources: constant voltage and constant current.

In a constant voltage supply, the output voltage is controlled with internal feedback to keep the output voltage constant regardless of the load. This means that even if the current draw changes due to a changing load resistance, the output voltage will stay the same.

This is the most common type of power supply. AC to DC converters are generally constant voltage supplies. A low drop out (LDO) regulator is a constant voltage supply. A battery is basically a constant voltage supply.

The output impedance of an ideal, constant voltage supply is very low. No matter the current, the output voltage change is nearly 0 V:

$$Z_{output} = \frac{\Delta V}{\Delta I} = \frac{\sim 0}{\Delta I} = 0$$

The second type of power supply is a constant current supply. In this type of supply, the output voltage is continually adjusted by internal feedback circuitry to keep the current coming out of the supply and through the external load constant. If the load resistance decreases, the output voltage decreases to keep the current through the load constant. If the load resistance increases, the output voltage increases to keep the output current constant.

Of course, there are limits to the max voltage that can be used with a constant current supply. If the load resistance increases so much that the supply can't provide a high-enough voltage to keep the current constant, the supply switches to constant voltage (CV) mode. But, if the current can be provided with an output voltage below the max voltage setting, the supply stays in the constant current (CC) mode.

The output impedance of an ideal constant current supply is surprisingly very high:

$$Z_{output} = \frac{\Delta V}{\Delta I} = \frac{\Delta V}{\sim 0} = \infty$$

When PCB traces are driven by a constant current supply, as was done in the experiments measuring the temperature rise of traces, there is a potential behavior that can arise which can result in the temperature of a trace increasing on its own, high enough to smoke, melt, and potentially act like a fuse and open. This effect is called *thermal runaway*.

The resistivity of copper is temperature dependent. The higher the temperature of the copper, the higher its resistivity. The figure of merit that describes this property is the temperature coefficient of resistance, α. For copper, it is about 0.4%/degC.

If the copper temperature rises, because the ambient temperature rises by 10 degC, the resistance of a copper interconnect will increase by 0.4%/degC x 10 degC = 4%.

The resistance of a copper trace, which includes the temperature dependence, is:

$$R = R_0 (1 + \alpha \Delta T)$$

5.6 Thermal Runaway with Constant Current

Where:

R = the resistance at any temperature above ambient

R_0 = the resistance at ambient temperature

α = the temperature coefficient of resistance of copper

ΔT = the temperature above ambient

If a constant current supply is used to drive a current through the copper trace, as the temperature rises, the resistance increases but the current would be the same. The same current through a higher resistance means the power dissipation increases, which increases the trace temperature.

The increased temperature further increases the resistance, and the power consumption increases even more. If the current is over a threshold, the temperature will continue to rise indefinitely until the wire gets so hot as to melt, fuse, and open up: a thermal runaway.

This condition can be illustrated by adding the temperature dependence of the resistance to the above analysis:

$$\frac{\Delta T}{\theta} = I^2 R = I^2 R_0 (1 + \alpha \Delta T) = I^2 R_0 + I^2 R_0 \alpha \Delta T$$

After a little algebra, the temperature rise over ambient is:

$$\Delta T = \frac{\theta I^2 R_0}{1 - \theta I^2 R_0 \alpha}$$

This says that when the current increases, as $\Theta I^2 R_0 \alpha$ approaches 1, the denominator gets smaller and the temperature difference explodes. At this point, the temperature will rapidly increase. This is thermal runaway.

The condition for the current required to instigate a thermal runaway is when the denominator explodes, or,

$$1 - \theta I^2 R_0 \alpha = 0 \quad \text{or} \quad I = \sqrt{\frac{1}{\theta R_0 \alpha}}$$

Small features of the thermal environment of the board that affects the thermal resistance can have a large impact on the temperature rise and the current when thermal runaway begins.

This behavior is easy to observe. When a constant current supply is used to drive a fixed current through a copper trace, the voltage across the trace is directly related to the instantaneous resistance:

$$V = I \times R$$

Since I is held constant by the power supply, the voltage is a direct measure of the resistance. As the temperature increases, so will the resistance and the measured voltage across the trace.

To measure the thermal runaway effect, the 6 mil trace was driven by a constant current power supply while the voltage across the trace was measured with a scope. This experimental set up is shown in **Figure 5.5**.

5.6 Thermal Runaway with Constant Current

Figure 5.5 Measurement system to apply a constant current to a 1-inch ling trace and measure the voltage across it with a 10x scope probe.

Below a current threshold, the resistance of the trace did not change as the current changed. The interconnect did not heat up and the resistance was constant.

When enough current was passed through the trace to heat it above ambient, the temperature rose to a higher equilibrium value, but then remained constant. When the wire was touched, the temperature momentarily dropped, the resistance dropped, and the voltage across the trace decreased. **Figure 5.6** shows an example of the measured voltage across the 6 mil wide trace with 3 A DC, constant current. The trace was hot to the touch.

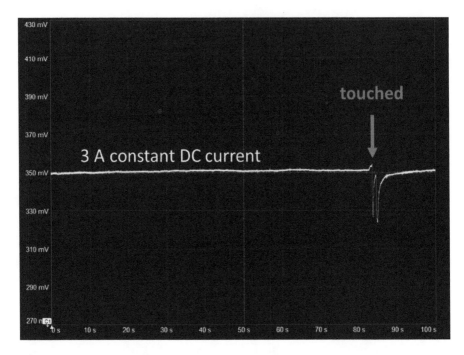

Figure 5.6 Measured voltage across the trace with 3 A constant DC current. The voltage was stable for 80 seconds and the wire was touched, decreasing its temperature, its resistance, and the measured voltage.

Even a current of 3.5 A through the 6 mil wide trace, while making the trace hot to the touch, did not start a thermal runaway. However, when the current was raised to 3.9 A, the thermal runaway condition was met. The temperature increased continuously until the trace got red hot, smoked, melted, and then fused open. **Figure 5.7** shows the measured voltage that is proportional to the resistance which is proportional to the trace's temperature and the trace's power dissipation under constant current.

5.6 Thermal Runaway with Constant Current

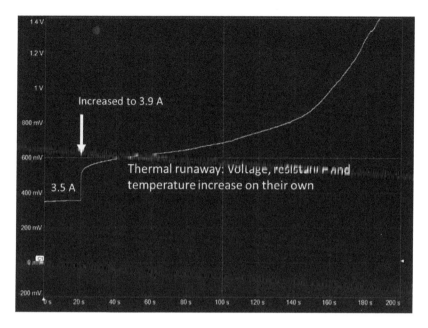

Figure 5.7 A DC current of 3.9 A through the 6 mil wide trace initiates thermal runaway and the temperature increases for almost 3 minutes before it fuses open.

This is a danger in constant current circuits with conductors having a positive temperature coefficient of resistance. If the current is above a threshold value related to the thermal resistance to the environment, it is possible for the temperature to increase on its own to a catastrophic end.

This behavior is not possible with a constant voltage supply. In this case, as the temperature of the trace increases and the resistance increases at constant voltage, the current will actually decrease and the power consumption will decrease. The power consumption self-limits at constant voltage.

While it is possible to melt a trace using a constant voltage supply by raising the voltage so the current through the traces exceeds 4 A, it may require small increases to the output voltage. As the trace heats up, it will decrease the current and the power consumption and self-limit the temperature rise.

Watch this video and I will show the maximum current through a narrow trace and demonstrate thermal runaway.

5.7 Practice Questions

1. What is the maximum current you can send through a 6 mil wide trace in 1 oz copper with a temperature rise of less than 40 degC?

2. What is the maximum current you can send through a 20 mil wide trace in 1 oz copper with a temperature rise of less than 40 degC?

3. What line width in 1 oz copper would you need to carry 10 A of current with a 40 degC temperature rise or less?

4. What happens to the resistance of a copper trace if the temperature of a trace increases?

5. What is the minimum trace width a low-cost fab shop such as PCBway can fabricate?

6. Why is it useful to use the minimum line width your fab shop can fabricate for a signal trace?

7. What would be a strong compelling reason to use a wide signal trace?

8. What would be a strong compelling reason to use thicker copper than 1 oz copper?

9. At what current would thermal runaway happen in a 6 mil wide trace in 1 oz copper?

10. If the copper trace were 2 oz copper, would the maximum current before thermal runaway started be higher or lower than for 1 oz copper?

11. What are two advantages of routing all power traces 20 mil wide?

12. What is the difference between a constant voltage and constant current power supply? How do you switch between the two modes of operation?

13. What is thermal runaway?

14. What is the type of power source that could cause thermal runaway?

Chapter 6
Trace Width Considerations: Series Resistance

The best design practice when selecting the trace width for signals is to use as narrow a trace width as practical to facilitate the routing of traces, especially between leads, via holes, and other component pads. This translates to a 6 mil wide trace as the default for signals and 20 mil wide for power traces.

The maximum current these traces can handle before there is any significant heating is 1 A for signal traces and 3 A for power traces.

The conductor line width influences the maximum current carrying capacity and also the series resistance. The series resistance is the simplest electrical property of a copper trace we can analyze.

In every application, you should consider the maximum acceptable resistance for an interconnect and estimate the worst-case series resistance of any interconnect. The difference between the requirement and the performance is the design margin. As long as the estimated resistance is well below the acceptable maximum value, the resistance of the interconnect is transparent and should not influence the performance of the circuit.

This recommendation applies to both solderless breadboard interconnects and printed circuit board traces.

6.1 Resistance of Any Uniform Conductor

In order to use a simple model to calculate the resistance of a conductor, we have to assume it has a constant cross section. This is referred to as a *uniform interconnect*. For example, if it is a long rod or a long strip, we can easily estimate the series resistance. These two examples of conductors with uniform cross sections are shown in **Figure 6.1**.

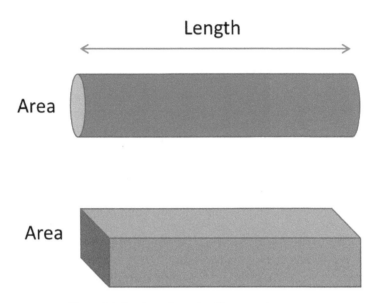

Figure 6.1 Cross sections of uniform conductors

As long as the cross section is constant down the length of the conductor, we can estimate the series resistance from one end of the conductor to the other, assuming a uniform current density across the face of the conductor. This is usually a good assumption especially at DC.

When the cross section is not uniform, such as when the diameter of the rod changes, or the shape of the rectangular strip changes, we can still divide the conductor into small, uniform sections, estimate the series resistance of each one, and add them up.

If we need a more accurate estimate, we would need to use a 3D field solver, such as Ansys Q3D, to calculate the series resistance for any arbitrary 3D structure.

When the cross section is uniform, the resistance from one face to the other can be estimated using the simple relationship:

6.1 Resistance of Any Uniform Conductor

$$R = \rho \frac{Len}{A}$$

Where:

R = the series resistance from one face to the other

ρ – the conductor's intrinsic bulk resistivity, 1.68 uOhm-cm = 0.66 uohm-in for copper, for example

Len = the length of the conductor from face to face

A = the cross-sectional area of each face that is uniform down the length

The resistance, R, depends on the three terms identified in this relationship: the bulk resistivity, a material property, the length, and the cross-sectional area.

The bulk resistivity will vary depending on the processing of the copper. The lowest resistivity possible is when it has been purified and is oxygen free copper (OFC). Then it is 1.68 uohm-cm. If it is then rolled into a sheet and annealed, it is 1.72 uohm-cm. When it is plated, depending on the plating chemistry and conditions, it can increase by as much as a factor of 2. This adds a little uncertainty to the resistance of traces.

For example, in a 24 AWG copper wire used to hook up components in a solderless breadboard, the radius is 10 mils. The resistance of a 1-inch long length is:

$$R = \rho \frac{Len}{A} = 0.66 [uohm-in] \frac{1[in]}{\pi (0.01[in])^2} = 2.1 mohm$$

Note that when using this relationship, it is very important to ALWAYS pay attention to the units and use a consistent set. This result says that the resistance of a 1-inch length of 24 AWG wire is about 2 mohms. This is a very good number to remember.

It is also very close to the specified series resistance. Most tables of wire resistances show the resistance of 24 AWG wire as 25 ohms/1000 ft = 2.0 mohm/in.

For example, a 10-inch length commonly used in hooking up interconnects has a resistance of 10 in x 2 mohm/in = 20 mohms.

The bulk resistivity is an intrinsic material property and does not depend on the shape or size of the piece of conductor. It is a measure of the ability of the material to carry current.

In addition to the bulk resistivity of the material, ρ, the bulk conductivity, σ, is often used to describe the electrical resistance of a conductor. The resistivity and conductivity are directly related as:

$$\sigma = \frac{1}{\rho}$$

For OFC copper, the bulk resistivity is 1.68 uohms-cm = 0.66 uohms-in. The conductivity is 59 /(uohms-m). The units of 1/ohms are also called Siemens. This makes the conductivity of copper 59 MegS/m.

The units of bulk conductivity or resistivity are unusual. They should not be interpreted as indicating a resistance of any specific length or shape of the material. Instead, they can be used to compare different materials. For the same geometry interconnect, the series resistance from one end to the other would increase with a larger bulk resistivity or smaller conductivity.

When using printed circuit board traces, rarely do we have the option of selecting other conductor materials than copper. However, it is still useful to be aware of the options so they are not a distraction, and for those cases where other conductor materials might be options.

Of course, keep in mind that it is not just the material that affects the resistance of a physical interconnect; the cross section and length of the conductor also play a dominant role.

6.1 Resistance of Any Uniform Conductor

For example, the best-case bulk resistivities of five popular conductor materials are:

Silver: 1.59 uohm-cm = 0.63 uohm-in

Copper: 1.68 uohm-cm = 0.66 uohm-in

Gold: 2.2 uohm-cm = 0.87 uohm-in

Aluminum: 2.65 uohm-cm = 1.04 uohm-in

Tin: 12.6 uohm-cm = 5.0 uohm-in

It is important to note that the lowest resistivity material is actually silver, not gold. The difference between silver and copper is very small, less than 10%. There is little electrical advantage to use silver conductors over copper and using silver is rarely worth the higher cost.

It is not true that gold is lower resistivity than copper. The resistivity of gold is about 30% higher than copper. The reason gold is so commonly used in interconnects is not because of its resistivity, but because it does not oxidize.

Most metals, when exposed to the air, form an oxide layer on their surface. The oxide is usually a higher resistance than the metal. This means two conductors with an oxide layer on their surfaces are brought in contact, they will have a layer of insulating material between them. This contributes to a sometimes very high contact resistance and a poor contact. This is the problem with a cold solder joint formed by a layer of tin oxide in the middle of a solder joint between two components. Tin oxide has a very high resistivity.

To prevent this high contact resistance when two metal surfaces are brought together, like in a connector, the metal surfaces are coated with gold, which does not oxidize. Two metal surfaces coated with gold will have a very low contact resistance and make a reliable contact over a long period of time, even when exposed to air.

6.2 Sheet Resistance of a Copper Layer

A trace on a board has a rectangular cross section. When it is fabricated from 1 oz copper, its thickness everywhere on the copper layer is 35 um or approximately 1.4 mils. An example of the geometrical features of a trace is shown in **Figure 6.2**.

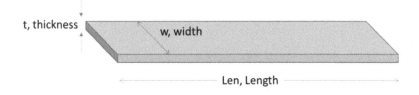

Figure 6.2 Geometric features of a PCB trace.

It is important to use the correct names for each of these features and the correct letters commonly used to describe their dimensions. While there is no rule that specifies what letter is used for what dimension, there are some conventions.

The thickness, t, of a trace refers to the dimension of the copper sheet in the vertical direction off the surface of the PCB substrate. This is based on the weight of the copper.

The width, w, of the trace is in the plane of the surface of the PCB and at right angles to the direction of current flow. Do not call the width of the trace the thickness of the trace. Thickness is reserved for the vertical dimension.

The length, Len, of the trace is measured in the direction in which the current flows from one end to the other. It is common to use the letter L or even d for the length of the trace. Avoid the use of the letter L as this can be confused with inductance. Avoid the letter d as this is ambiguous and is often used for a dielectric thickness. Using the letters Len is unambiguous.

For a uniform trace on a circuit board, we can estimate the series resistance of a trace from one end to the other end using the simple relationship:

6.2 Sheet Resistance of a Copper Layer

$$R = \rho \frac{Len}{A} = \rho \frac{Len}{t \times w} = \frac{\rho}{t} \frac{Len}{w} = R_{sq} \times n$$

Where

ρ = the bulk resistivity of copper = 1.68 uohm-cm

Len = the length of the trace

t = the thickness of the copper trace, for 1 oz copper, 34 u

w = the trace width

R_{sq} = the sheet resistance with units resistance per square

n = number of squares down the trace

The last version of this relationship is incredibly powerful. It will dramatically simplify how we calculate the resistance of any trace etched from a copper layer.

The trace thickness will be exactly the same for every trace etched from the same copper layer. This means the ratio, ρ/t, will be the same for every trace etched from the same layer. This figure of merit is given a special name, sheet resistance, Rsq.

The resistance of a trace is the (sheet resistance) x Len/w. If the trace is in the shape of a square, so that Len = w, the resistance from one edge to the other is literally just the sheet resistance. The trace can be 0.1 inches x 0.1 inches, or 10 inches x 10 inches. The ratio of Len/w is always the same.

This means that the resistance of any square-shaped trace etched from the same copper layer will have exactly the same resistance, equal to the sheet resistance. This is counterintuitive. How can a square 0.1 inches on a side have the same edge-to-edge resistance as a square 10 inches on a side?

If we double the length of a trace, the path length doubles and the resistance will double. But then if we also double the width to keep the square shape, the resistance will be cut in half since the cross-sectional area has doubled. These two-dimensional changes cancel out. Every square-shaped trace in the same sheet will have the same edge-to-edge resistance.

This property of every square cut from the same sheet having the same edge-to-edge resistance is a powerful principle that we will leverage to quickly estimate the resistance of traces.

The resistance of a square-shaped trace is a unique resistance which is a figure of merit for each layer of copper. We call the resistance of one square cut out of the same sheet the resistance per square, or the sheet resistance, Rsq.

This is the resistance from edge-to-edge of any square-shaped trace. This is why the units of sheet resistance are ohms/square.

For 1 oz copper, the sheet resistance is

$$R_{sq} = \frac{\rho}{t} = \frac{1.7 \times 10^{-8}\ \Omega - m}{34\ \text{um}} = 0.5\ m\Omega$$

For ½ oz copper, the thickness is half that of 1 oz copper and the sheet resistance is twice this value, 1 mohm/sq. This is an easy number to remember.

In a 1 oz copper layer, each square is 0.5 mOhms. To find the total resistance of any uniform trace, we just count how many squares are down its length. This is n = Len/w. Each square's resistance is in series so the total resistance of the trace is 0.5 mohm/sq x n.

A trace 10 mils wide and 1 inch long has 1 in/0.01 in = 100 squares down its length. Each square in a 1 oz copper sheet is 0.5 mohms/square, so the total resistance of the trace is 0.5 mohms/sq x 100 sq = 50 mOhms.

It really is as simple as that. Counting the squares in a trace is illustrated in **Figure 6.3**.

6.2 Sheet Resistance of a Copper Layer   151

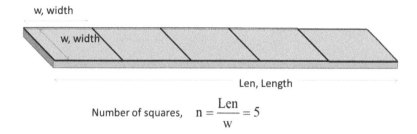

Figure 6.3 The number of squares on a trace is how many times the width fits down the length.

Just at a glance, it's possible to estimate the number of squares down a trace.

For a minimum width trace, 6 mil wide, the resistance for a 1-inch length is 0.5 mohm/sq x 1000 mils/6 mils = 0.5 x 170 = 85 mohm.

Knowing the resistance per length of a specific trace width is a handy figure of merit to remember. The resistance per length is just

$$R_{Len} = \frac{R}{Len} = \frac{R_{sq}}{w}$$

For the case of a 6 mil wide trace length, the resistance per inch is

$$R_{Len} = \frac{R_{sq}}{w} = \frac{0.5 m\Omega/sq}{0.006 in} = 83 m\Omega/in$$

Note that the units of squares in the sheet resistance is dimensionless and just disappears.

For a 20 mil wide trace, the resistance per length is just 0.5 mohm/sq /0.02 inches = 25 mohm/in.

We can use this estimate of 83 mohm/inch as a good measure of the resistance per inch of any signal trace. A trace 3 inches long would have a series resistance of 3 in x 83 mohm/in = 250 mohm of series resistance.

When calculating the number of squares, the most important consideration is to use a consistent set of units for the length and the width. The number of squares is dimensionless.

Get in the habit of looking at the traces on your circuit board and quickly estimating the resistance of each trace. This will give you a feel for one of their electrical properties and what to expect for their performance.

> *To quickly estimate the resistance of a trace on your board, use the rules of thumb: for 1 oz copper, the series resistance of a 6 mil wide trace is 83 mohms/inch and, for a 20 mil wide trace, it is 25 mohms/inch.*

> *Watch this video and I will walk you through the concept of sheet resistance and why it is such a useful term to characterize copper sheets and to estimate trace resistance.*

6.3 Measuring Very Low Resistances

The typical resistance of traces, even the narrowest traces, are generally much lower than 1 ohm. This low a resistance presents some measurement challenges.

It is easy to measure a resistance with a DMM set as an ohmmeter, but it is sometimes difficult to measure a resistance lower than 1 ohm without introducing an artifact. There are three common problems that prevent accurate low-resistance measurements with a 2-wire ohmmeter:

- ✓ The internal offset voltage of the DMM used to measure the current through the device under test (DUT) or the voltage across it
- ✓ The series resistance of the leads
- ✓ The contact resistance at the interface

6.3 Measuring Very Low Resistances

In some ohmmeters, it is possible to compensate the DMM for those offsets that are constant, such as the internal offset voltage or the lead resistance, but the contact resistance may vary from measurement to measurement.

You can demonstrate this for yourself by taking your DMM, setting it to the lowest resistance scale and measuring the resistance with the leads shorted. **Figure 6.4** shows examples of two different DMMs with two different types of shorted leads, showing a measured resistance of 0.21 ohms for short leads and 1.5 ohms for the standard clip leads.

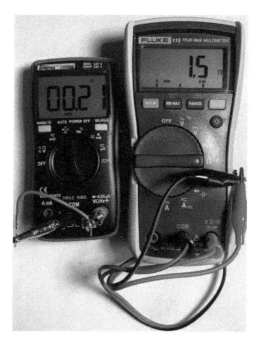

Figure 6.4 Examples of the shorted resistance of two common DMM set as ohmmeters. This means measuring resistances lower than 1 ohm will be difficult.

The way around these common problems is to use a technique developed in 1861 by William Thompson, also known as Lord Kelvin, referred to as the Kelvin 4-wire method.

154　Practical Guide to Prototype Breadboard and PCB Design

The general method of measuring the resistance of a component is to force a current through it and measure the voltage across it, V, and the current through it, I. The resistance of the component is R = V/I.

In the conventional 2-wire method, there are two connections to the DUT. The current flows through the leads and the voltage is measured at the ends of the leads. This voltage measurement includes the series resistance of the wire leads and the contact resistance of the leads to the DUT at each end.

The innovation Lord Kelvin introduced was to separate the leads and contacts that measured the voltage across the DUT from the forced current. The lead resistance and contact resistance is still present in the circuit with the forced current.

But the voltage measurement just includes the voltage drop across the DUT, not across the lead or contact resistances. Any lead resistance or contact resistance in the voltage measurement path has no effect on the voltage measurement. These two measurement topologies are illustrated in **Figure 6.5**.

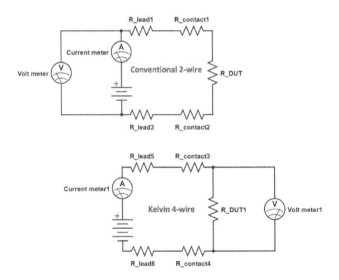

Figure 6.5 The measurement topology for the conventional 2-wire method (top) and the 4-wire method (bottom).

To implement the 4-wire method, it is necessary to attach one set of leads to the DUT to force the current and a different set of connections to measure the voltage induced across the DUT from the current. The resistance of the DUT independent of the contact resistance and lead resistance is just R = V/I.

For example, to measure the resistance of the vertical column in a solderless breadboard, we use one set of leads to force the current by connecting to a power supply. We add a series ammeter to measure this current. Then we connect a separate pair of leads to the voltmeter, making separate connections to the vertical column strip. This configuration is shown in **Figure 6.6**. In this case, we measure a voltage of 37.1 mV with a current of 1.000 A. This is a resistance of

$$R = \frac{V}{I} = \frac{37.1 \text{mV}}{1.000 \text{A}} = 37.1 \text{ mohms}$$

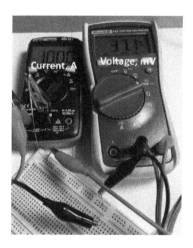

Figure 6.6 Measuring the resistance of half a vertical column in a solderless breadboard. One pair of leads connects to a power supply and forces 1.000 A of current through the column. Another pair of leads connects the voltmeter and measures a voltage drop of 37.1 mV.

This is the resistance of half a vertical column in a solderless breadboard.

When I connected the voltmeter leads to the wires driving the current through the column, rather than making a separate connection to the column, I measured a voltage of 90 mV. This extra voltage is the voltage drop across the contact resistance of the wire inserted into the hole of the solderless breadboard. The difference is about 50 mV, or 25 mV for each contact. This is a contact resistance of R = V/I = 25 mV/1 A = 25 mohms. This is a rough measure of the contact resistance of a wire plugged into a solderless breadboard.

Using the conventional 2-wire method, I measured a resistance of 1.9 ohms as the resistance of the column in the solderless breadboard. We now see most of this resistance is artifact, the lead resistance of the DMM wires.

If we use a forcing current of 1 A and our DMM is capable of measuring a voltage as low as 1 mV, we can measure a resistance as low as R = V/I = 1 mV/1 A = 1 mohm. By pushing the smallest voltage to 0.1 mV, we can routinely measure resistances as low as 100 uohms.

The kelvin 4-wire method is a very powerful technique to measure sub-mohm resistances. Whenever it is necessary to measure a resistance below 1 ohm, the 4-wire method should be used.

Watch this video and I will show you how to measure a trace with the 4-wire method.

6.4 Voltage Drop Across Traces

If the typical maximum length of a trace on your board is about 5 inches, then for 1 oz copper and 6 mil wide trace, the typical maximum resistance would be about 83 mohm/inch x 5 in ~ 0.4 ohms. For a 20 mil wide trace, the maximum resistance might be 25 mohm/in x 5 in = 0.125 ohms.

Is this resistance a lot or a little? How much resistance is too much?

There are two criteria to judge the relative importance of the series resistance of a trace:

- ✓ How much voltage drop there is in the signal trace due to the IR drop
- ✓ How large the trace resistance is compared to the resistance of the voltage source.

For a signal trace and a maximum resistance of 0.40 ohms, the voltage drop with a current of 100 mA would be 0.040 V or 40 mV. In most applications, this voltage drop in a signal path carrying 100 mA may not be significant. When connecting an analog signal, in which case small voltage drops may be significant, the current might be on the order of 1 mA and the voltage drop would be as low as 0.4 mV.

For a power path with a trace series resistance of 0.125 ohms, the maximum current we might see, limited by potential trace heating, is 3 A. This would result in a voltage drop of as much as 3 A x 0.125 ohms = 0.375 V. This can be a significant voltage drop in some situations. For a 5 V rail, a 0.375 V drop to the device being powered is a 7.5% drop.

In very high current applications, and when small voltage drops are important, it may be necessary to use a trace wider than 20 mils for delivering power. This is why it is so important to get in the habit of putting in the numbers to estimate performance and doing your own analysis.

The second criteria to use to evaluate how much resistance is too much is in comparison to the resistance of the source that drives the current.

6.5 The Thevenin Model of a Voltage Source

The absolute simplest ideal equivalent circuit model for a power source is an ideal voltage source. The fundamental property of an ideal voltage source is that its output voltage is constant, no matter the load attached. This model matches the behavior of many real-world power sources when they supply very low currents.

A far more accurate equivalent circuit model that takes into account the behavior of a real power supply to higher current loads is a Thevenin circuit model. This is a powerful model and should always be the starting place to describe a real power supply.

Every voltage source, either from a microcontroller's digital output pin or from a voltage regulator module (VRM) such as a switch mode power supply (SMPS), a low drop out (LDO) regulator, and even a battery, can all be described to first order as an ideal Thevenin voltage source.

This means the voltage source can be described by an equivalent circuit model composed of an ideal voltage source and an ideal series resistor. This equivalent circuit is shown in **Figure 6.7**.

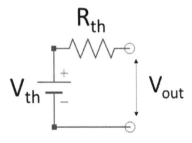

Figure 6.7 Basic Thevenin model of a voltage supply.

An ideal voltage source will always keep its output voltage constant, no matter the current load. This means that the impedance looking into an ideal voltage source is 0 ohms: $Z = \Delta V/\Delta I = 0/\text{anything} = 0$.

The series resistor in the circuit means that as the current from the source increases, the voltage at the output of the power source will decrease due to the internal voltage drop across the Thevenin resistor.

Only two terms completely characterize a Thevenin voltage source:

6.5 The Thevenin Model of a Voltage Source

- ✓ The Thevenin voltage, V_{th}
- ✓ The Thevenin resistance, R_{th}

These two terms are the important figures of merit of any power supply. Knowing their values will tell us about the performance of the power supply under many typical applications.

Unfortunately, however valuable this model and these two figures of merit are in describing a voltage source and using it effectively, rarely are these terms directly specified in the datasheet of a component. This is one reason it is important to be able to *reverse engineer* these figures of merit by measuring them.

Reverse engineering is a very valuable skill we will use over and over again. It is a process that extracts the behavior and properties of a component or system based on an assumed model.

Using our best guess, intuition, or other insight, we create an ideal model of the device. We perform measurements on the device and compare the measurements to the simulations of the model. We adjust the parameters of the model until the simulated voltages or currents match the measurement.

When we get good agreement, we have confidence the model is a good approximation of the real component, and the parameter values we used become the figures of merit for the device.

To apply reverse engineering principles to a power source, we assume the equivalent circuit model of the actual VRM DUT is the ideal Thevenin voltage source model. The voltage on the output of the voltage supply, V_{out}, with no load is a measure of the ideal Thevenin voltage parameter. This is the easy part.

In principle, the way to measure the Thevenin resistance parameter is to short the end of the voltage source and measure the current through the short. The Thevenin resistance is the ratio of the Thevenin voltage to the short circuit current. In practice this is never a good idea.

The Thevenin model is only a first-order model to describe the real voltage supply. It may apply well when the current draw is small, but this simple model may not match the device behavior when the current load is large and approaches the short circuit current. Many voltage sources have a current clamp or have a nonlinear transistor output stage. This means when the output of the voltage source is shorted, the device may be operating in a completely different mode than when the output is not shorted, and our first-order model may not be a good approximation.

A practical approach to extract the equivalent ideal Thevenin resistance is to add a resistive load to the source that is comparable to or a little larger than the internal Thevenin resistance. This forms a voltage divider. We should choose a load resistance so that the voltage drop with the load is not larger than half the Thevenin voltage of the source.

The output voltage across the load is a direct measure of the Thevenin resistance. **Figure 6.8** shows the equivalent circuit.

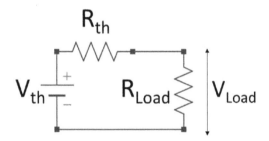

Figure 6.8 Circuit used to measure the loaded voltage of a Thevenin circuit.

The extracted Thevenin resistance is:

$$R_{th} = R_{load}\left(\frac{V_{th} - V_{load}}{V_{load}}\right)$$

The typical Thevenin resistance of a simple AC to DC converter or a battery or a signal source is on the order of 0.2-10 ohms. A 50 ohm load is a convenient value resistive load to add to reverse engineer the Thevenin resistance.

Many AC to DC power supplies, for example, show a current rating along with their DC voltage output. This current rating is very misleading. What limits the current? If the current rating is 1 A, does this mean that if the external load is low enough to draw 1.1 A, the power supply will not work or will turn off, or will explode? Will the 1 A-rated power supply never provide 1.1 A of current?

The current rating says nothing about the output Thevenin resistance. It is usually based on the maximum power the supply can provide with acceptable temperature increase due to the power dissipation handling of the packaging. This is the maximum current the power supply can handle without any long-term thermal issues. It is not about the maximum current the supply can deliver under shorted load or related to the Thevenin resistance.

A simple way of reverse engineering the Thevenin resistance of a voltage source is to measure the output voltage of the source open circuit and then again after a resistive load is connected across the output of the power source. A scope or DMM can be used to measure the output voltage of the power source with and without the external resistor attached.

The resistor attached as the load should be low enough to cause an easily measured voltage drop when loaded, but not a voltage drop larger than half the unloaded voltage. A good initial resistance to try is a value on the order of 50-100 ohms.

As a simple alternative, an oscilloscope can be used to perform this measurement, taking advantage of its built-in 50 ohm input resistance. To use the internal 50 ohm resistance of most scopes, it is important to use a direct coax cable connection to the scope and NOT a 10x probe. The voltage measured on the output of the source under this no-load condition, with a 1 Meg input to the scope, is the Thevenin voltage.

Then the scope is set for a 50 ohm input resistance. Just be sure the voltage on the output of the source is not larger than 5 V. There is a limit to 5 V as the largest RMS voltage that should be connected to any scope set for 50 ohms due to the power dissipation ability of the scope's internal 50 ohm resistor. It can only consume 0.5 watts before it may be damaged. If an input rms voltage of more than 5 V is applied to the 50 ohm input resistance of the scope, the internal resistor and circuit board to which it is mounted may be thermally damaged.

The measured voltage drop on the supply output when the 50 ohm scope resistance is applied to the device can be used to extract the Thevenin resistance.

This technique was used to measure the 5 V rail from an Arduino Uno powered by an external USB hub. The Thevenin voltage, measured with a 1 Meg termination in the scope, was 5.15 V. When the scope was set for a 50 ohm load, the voltage measured on the 5 V rail was 4.93 V. The Thevenin resistance is

$$R_{th} = R_{load} \left(\frac{V_{th} - V_{load}}{V_{load}} \right) = 50\Omega \left(\frac{5.15V - 4.93V}{4.93V} \right) = 2.23\Omega$$

There is no law that says a real voltage source must match the behavior of this first-order ideal circuit model. However, it is remarkable that many real voltage sources do show an output voltage drop with current that matches the behavior predicted by this simple ideal circuit model.

The fact that real voltage sources really do behave as this first-order Thevenin model predicts is what makes this model so valuable in describing real sources. Using this model, the extracted figures of merit of the Thevenin voltage and resistance become valuable figures of merit to describe any real voltage source.

6.5 The Thevenin Model of a Voltage Source

Watch this video and I will walk you through this simple process of reverse engineering the Thevenin voltage and resistance of two power sources.

A simple circuit can be used to routinely measure the equivalent Thevenin voltage and resistance for any voltage supply using an Arduino microcontroller board as an automated measurement tool. An example of this circuit is shown in **Figure 6.9**.

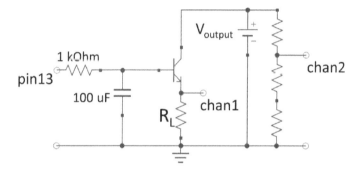

Figure 6.9 Circuit used to automatically measure the output voltage on chan2 of the Arduino, and the current load on chan1 of an Arduino. This will automatically extract the figures of merit of a Thevenin circuit model.

A digital output pin of an Arduino generates a slowly increasing voltage using an RC low-pass filter. This voltage ramp drives a transistor current source. As the current load to the power supply increases, the voltage across the source and the current through it are measured by two of the analog to digital converter (ADC) input channels of an Arduino used as a measurement instrument.

From the measured voltage drop on the rail and the measured current from the source, the open circuit voltage and the Thevenin resistance can be extracted. These are the parameters or figures of merit of the equivalent ideal circuit model of the voltage source.

For example, the output voltage of a 100 mA rated, 12 V AC to DC convertor was measured as the DC current load changed. The slope of the V vs I curve is a direct measure of the output resistance. **Figure 6.10** shows the measured data with a slope that is very constant with a value of 7.2 ohms. This is the measured Thevenin output resistance of the power supply.

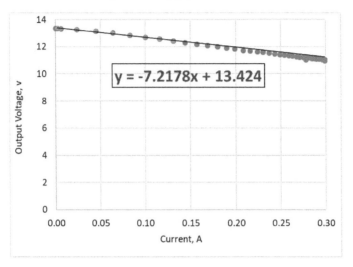

Figure 6.10 Measured output voltage as the current load increases on a 12 V AC to DC converter.

In this example, the slope is very constant, but above 0.2 A current load, the output voltage drops faster than just a simple 7.2 ohm resistor would predict. This measurement shows when the simple first-order model begins to break down and a more complicated behavior begins to appear. However, even up to a current of 300 mA, the Thevenin model is still a very good approximation.

Not all voltage sources match this simple Thevenin model, especially as the current load increases. Many devices, especially signal sources, have a current clamp in their outputs making the output resistance very nonlinear. The Thevenin model is only a rough approximation within a limited current range.

For example, a digital output pin of an Atmega 328 microcontroller, the heart of an Arduino Uno, shows this nonlinear behavior. **Figure 6.11** is the measured V-I curve and the resulting extracted Thevenin resistance as the current load to an Arduino digital I/O pin was increased. The match to an ideal Thevenin model, which would have predicted a constant output resistance, is not a very good fit.

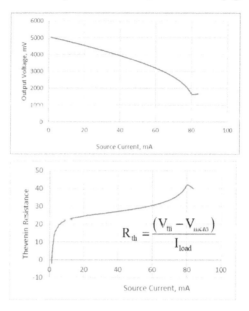

Figure 6.11 The measured output voltage of the digital I/O pin of an Arduino. Top: the output voltage as the current load increased. Bottom: the extracted Thevenin resistance at different current loads.

This shows that in the range of 3 mA to 60 mA of load current, the output resistance of a digital pin is about 25 ohms. This is a good estimate to use for the output resistance of an Arduino digital pin. For other microcontrollers, this resistance can be much higher.

6.6 How Much Trace Resistance Is too Much?

The second criteria for which to decide if the trace resistance is too high is how it compares to the Thevenin resistance, which is the source resistance of whatever is driving signals on the trace. To be transparent, the trace series resistance should be much smaller than the signal source resistance. This means it is important to know the source resistance of your voltage source.

Where possible, you should search for information about the output resistance of all the sources used in your circuit. Rarely is this information offered in datasheets. Sometimes you will have to reverse engineer the Thevenin voltage and Thevenin resistance of your sources. Sometimes you will have to guess a worst-case rough estimate.

This is why building a small eval board to characterize your devices can be so valuable.

If you can't find the information or are not in a position to measure it, assume that for a power rail, the output resistance is 1 ohm. For a signal source, the output resistance is 25 ohms. If knowing an accurate value is important, then you must measure the output resistance of a sample.

These values are to be compared to the series resistance of the traces connecting the voltage source to the DUT.

When the trace resistance is very small (< 20%) compared to the source resistance, the impact of the voltage drop in the trace is insignificant and not important.

When the worst-case series resistance for a 5-inch signal trace is on the order of 0.4 ohms, its impact is typically less than 0.4 ohms/10 ohms = 4%. Its resistance is insignificant *compared to* the output resistance of the source.

For a voltage regulator module (VRM), the output resistance might be on the order of 1 ohm. The worst-case trace resistance for a 5 inch long, 20 mil wide trace is 0.125 ohms. This is less than 0.125 ohms/1 ohm = 12% of the output resistance of a typical VRM and would generally be insignificant. Of course, it would only show up as a voltage drop when there is high current flowing in the power trace.

If the trace resistance of a power path is found to be too high compared to the output resistance of the VRM for a specific application, this is a compelling reason to use a wider trace width than 20 mils.

6.7 The Resistance of a Via

In a multilayer printed circuit board, connections are made between traces on one layer to another layer using structures called vias. The most common type of via is a through-hole via also referred to as a plated through-hole (PTH) via. A cross section of a PTH in a 4-layer board is shown in **Figure 6.12**.

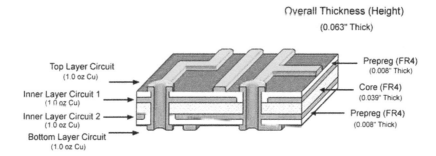

Figure 6.12 Example of a 4-layer board with through-holes, courtesy of Altium.

The PTH vias are added to the board after all the layers have been laminated together and before any of the top layers are patterned. At the location of each via, a hole is drilled through the board. Any traces on any inner layers the hole cuts through leaves the edge of the trace exposed at the edge of the hole.

If the via passes through a plane on a layer where there should be no connection, a clearance hole, or antipad, is etched in the plane where the drill will pass through.

After drilling, the board is dipped into a catalyst solution that coats the inside of the holes and grows a thin layer of copper.

The board is then dipped into another bath where this thin layer of copper and the entire exposed top layer is electroplated with a 1 oz thick copper layer. The copper on the surface of the board is plated up and the inside of the via holes are plated with about this thickness of copper.

Any traces that were cut by the drill with their edges exposed inside the hole are now electrically connected together with pads on the top and bottom surfaces of the board. An example of the cross section of a plated through-hole via is shown in **Figure 6.13**.

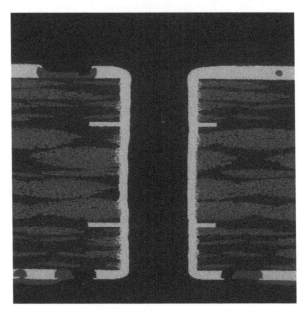

Figure 6.13 A cross section in a 4-layer board showing the PTH with pads on the top and connecting to traces on inner layers. Image source SEM Lab, inc.

When adding a via to a board to connect two or more layers, there are generally three important design features to select:

1. The drill hole size

2. The outer diameter of the annulus of the capture pad on the top and the bottom layers

3. The outer diameter of the clearance hole in a plane the via passes through and to which it is not connected.

With so many possible values to choose from, what should be the best design practices for selecting via feature dimensions?

Just as with the selection of the optimized line width, the via features should be chosen for the smallest size to optimize routing, at no cost adder, and without sacrificing electrical performance.

The starting place is to use the minimum features the fab shop is capable of producing without extra cost. Then, evaluate the impact on current handling and series resistance. If the DC performance is acceptable, use this minimum, lowest-cost size.

Most low-cost fab shops can drill a hole as small as about 0.3 mm. This is 11.8 mils. To be safe and not have to worry about upcharges for too small a via, a good default value for a PTH drill size is 13 mils.

Since the via hole is plated with copper, there is a difference in hole diameter between the drilled hole and the plated hole. This difference is 2x the plating thickness. We refer to the via hole diameter before plating as the drill diameter, while after plating, the hole diameter is the finished hole size (FHS). This will be narrower by 2x of the wall-plated thickness.

If the drill diameter is 13 mils and the plating thickness is 1.3 mils, the FHS will be 13 mils − 2 x 1.3 mils = 10.4 mils. This is not important if the via is used for signal connection but can be very important if a through-hole lead is to be inserted into the hole. In this case, a decrease in hole size of 2.6 mils can be the difference between a good fit and an impossible fit.

When designing for through-hole components, it is important to use the FHS as the design parameter, keeping within the constraints of the fab shop.

The typical minimum line width for most low-cost fab shops is 6 mil for line or space. This is also the minimum width annulus or ring of copper around a plated through-hole. With a 13 mil diameter drilled hole and 6 mil annulus capture pad, this is an outer diameter of the capture pad of 13 mils + 6 mils + 6 mil = 25 mils.

These are the best recommended practices to use as the default value of any via connecting signals on different layers: 13 mil drill hole diameter and 25 mil diameter capture pad. This setup is shown in **Figure 6.14**.

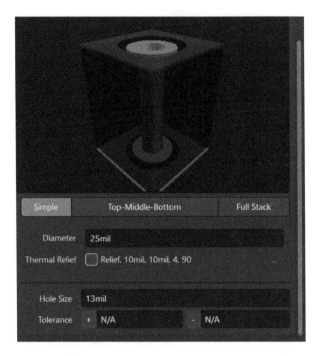

Figure 6.14 An example of setting up the features of a via in Altium Designer.

In some PCB EDA tools, it is confusing which term is which. Always look for a dimension referred to as drill size or drill hole. This should be set for 13 mils. Another feature, labeled as just a diameter, is often the outer diameter of the capture pad on the top and bottom layers. This should be set for 25 mils.

The advantage of using the smallest via at no extra cost is that it will take up the least space and provide the highest density and easiest routing.

The downside of using a small diameter via is that it will have a higher resistance than a larger size via. How much resistance is there in a via? Before we go through the analysis, take a moment to write down your guess of the resistance of a 13 mil diameter via, from one end to the other, through a 62 mil thick circuit board plated with 1 oz copper.

6.7 The Resistance of a Via

The resistance of a via can be calculated as a cylindrical annulus with a cross-sectional area and a length. This is straightforward.

However, there is another way which gives insight to the power of the sheet resistance concept and just counting squares.

Imagine we slit the via down its side and unroll it and lay it flat. The plated copper inside of the via will unroll as a rectangular piece of copper that has a length equal to the thickness of the board, 62 mils, and a width equal to the circumference, 3.1416 x 13 mils = 41 mils. This is illustrated in **Figure 6.15**.

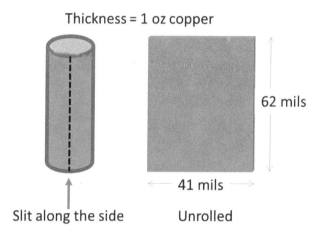

Figure 6.15 Example of slitting and unrolling a via to turn it into a rectangle of copper.

Now it is trivial to calculate the series resistance of the via. The thickness of the copper is 1 oz. In typical copper foil, the sheet resistance is 0.5 mohm/sq. When a copper layer is plated up, the bulk resistivity of the copper is a little higher than foil, but without the specific details for the plating process, this estimate of 0.5 mohm/sq is a good starting place.

The number of squares down the length of the via is 62 mils/41 mils = 1.5. This means the series resistance of a via is about 0.5 mohm/sq x 1.5 squares = 0.75 mohms. This is less than 1 mohm series resistance! How close did you get with your initial guess?

This simple analysis suggests that the series resistance of even the narrowest via will be very small and negligible compared to the typical trace resistance in any normal operation.

The effective width of the via when unrolled is 41 mils wide. It can easily handle the maximum 3 A of a 20 mil wide trace.

A 13 mils drill diameter via should always be the default via diameter to use unless you have a strong compelling reason otherwise.

6.8 Resistance of a Thermal Relief Via

When a via connects to a plane, there is one additional problem to watch for. A plane of copper is a very good conductor of heat. When there is a direct connection between a via and a plane, the plane will suck thermal power from the via with a low thermal resistance.

This low thermal resistance path from the via to the plane will make it extra difficult to solder a component to a pad with a via connecting to a plane. If the via hole is used to insert a through-hole pin and the via connects to a plane, the heat will be sucked into the plane before the pin can heat up to reflow the solder. This makes it very difficult to solder a through-hole pin when the via is also connected to a plane.

The way around this problem is to add a structure to the via that will increase the thermal resistance between the via and the plane. This is called a *thermal relief*. An example of a thermal relief structure in a via is shown in **Figure 6.16**. This is sometimes referred to as an isothermal pad.

6.8 Resistance of a Thermal Relief Via

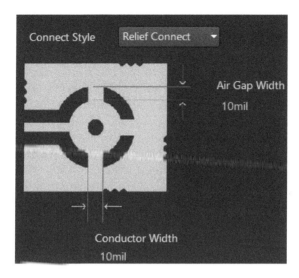

Figure 6.16 Example of a thermal relief structure added to a via to increase the thermal resistance between the via and the pad or plane to which it connects.

Thermal isolation is accomplished by having a moat of copper-free region to act as thermal isolation. To provide the electrical connection across that moat, a narrow drawbridge of copper is added between the via's capture pad and the plane or pad to which it connects. This provides some electrical connectivity, but high thermal resistance. The typical dimensions for the connections between the via capture pad and plane are 10 mils wide and 10 mils long. They could go as low as 6 mil wide and 6 mil long.

Once again, it looks like these small drawbridges would introduce high resistance to the signal path. After all, they are very narrow. Take a moment to jot down your guess for the added electrical resistance of the thermal relief structures in a via.

Using the principles of sheet resistance and counting squares, we can instantly estimate the series resistance of the connection through the drawbridges from the via to the plane.

Each drawbridge, or tab, is literally in the shape of a one square, 10 mils wide by 10 mils long. The resistance of each tab is 0.5 mohm/sq x 1 sq = 0.5 mohms. There are four of these tabs in parallel between the via and the plane. The total resistance is just ¼ x 0.5 mohm =

0.12 mohm. There are two sets of these thermal relief structures in the top and bottom layers of the board, so the total extra resistance of the thermal relief via is about 0.25 mohms.

The extra electrical resistance of a thermal via is completely negligible to the already very small resistance of a via and can be ignored. However, this structure provides a high-enough thermal resistance to make it easy to solder a pin to the via capture pad.

It is important to note that the only purpose of a thermal relief structure is to provide thermal isolation between the via and the pad to which it connects. If there is no thermal impact from a via, do not use a thermal relief structure, use a solid connection between the via and the pad or plane to which it connects. This will enable the smallest footprint for the via.

6.9 Practice Questions

1. Why do you want to use the narrowest line width practical?
2. What are two criteria that determine how narrow a line you should use?
3. What is the sheet resistance of 1 oz copper?
4. How many squares are in a 6 mil wide trace, 1 inch long?
5. What is the resistance of a 6 mil wide trace, 1 inch long in 1 oz copper? In ½ oz copper?
6. What is the minimum via drill hole for no extra cost?
7. What is the minimum line width and space for no extra cost?
8. What is the difference between the trace width and the trace thickness?
9. How does the bulk resistivity of the copper change as the trace width is increased?

10. How does the sheet resistance of the copper sheet change as the trace width is increased?

11. How does the resistance per length of a trace change as the trace width is increased?

12. The sheet resistance of 1 oz copper is 0.5 mohms/sq. If you measure the edge-to-edge resistance of a piece of copper 100 mils on a side, or 10 inches on a side, what will its resistance be?

13. A trace 10 mils wide is 2 inches long. How many squares are down its length? What is the resistance of the line if it is made with 1 oz copper?

14. A trace is 20 mils wide. What is the resistance of a 1-inch trace made with 1 oz copper? What is the resistance of a 20-mil wide trace 10 inches long?

15. How much resistance could a signal trace have and not affect most signals?

16. What is the max current you can run through a 6-mil wide trace before the temperature just barely increases?

17. What is the maximum current you can run through a 20-mil wide trace before the temperature just barely increases?

18. What is the resistance of a via?

19. How much current can you safely put though a 13 mil diameter via?

20. How much electrical resistance is there in a thermal relief via?

21. What is the difference between a thermal relief via and a thermal via?

Chapter 7
The Seven Steps in Creating a PCB

In every project, there are seven steps we will follow to go from the BoN sketch to a functioning widget. Some steps are more important than others.

No rules can guarantee success. But following these steps and the best practices can help you stack the deck and increase your chance of success and decrease the risk of a known problem arising.

You may be the first engineer to build the specific widget you are designing, but you are not the first engineer to build a working widget. Many before you have walked this path and encountered many problems. These are well documented.

While it is important to learn from your mistakes, it is equally important to learn from others' mistakes, so you do not have to make their mistakes over again.

> *A best design practice is a guideline to help you avoid a problem another engineer encountered and documented.*

For example, "add silk screen markings on your board to identify what the flashing LED is indicating" is a best design practice to avoid the confusion — read risk — of not seeing a potential fault or the lack of power to a circuit.

Never blindly follow any design guideline unless you understand what purpose it serves and if it applies to your specific design. You may find in your application you have a strong compelling reason not to follow it. Maybe in your design you want to keep some proprietary secrets and not identity you have an "error detected" LED indicator on your board.

The seven steps and their milestones for the completion of each step in every project, introduced in section 1.7, are again reiterated here and explored in more detail in this chapter:

1. Completed the plan of record (POR).
2. Completed the initial bill of materials (BOM).
3. Completed the schematic capture and final BOM.
4. Completed the layout and order all the parts.
5. Completed the assembly.
6. Completed the bring up, troubleshoot, and final test.
7. Completed the documentation.

In the following sections are check lists to verify you have completed the major steps necessary to finish each phase and are ready to move to the next phase.

The focus is on introducing and establishing:

- ✓ Best design practices
- ✓ Best assembly practices
- ✓ Best measurement practices
- ✓ Best analysis practices.

7.1 Step 1: Plan of Record

The purpose of the POR is to force you to think through the entire project at the beginning before major resources are committed. This is the chance to consider the big-picture trade-offs between the performance you can attain and the risk, cost, and schedule.

*The more risk sites you can identify and either design out, or
plan backups for, the higher your chance of success.*

If you tackle your project serially, completing one task, then thinking about the next task, then the next, and the next, your project will either fail, take too long, or you chose too simple a project, and someone has already made it before you.

The more details you can include in the POR, the more you will be forced to think through your entire project early in the design cycle.

The POR is a living document. As you learn more about your project, and some questions are answered, keep the POR updated. It may change tomorrow, but at least today you have a vision of the path you will take and where you will end up when you are done.

Be sure to include the following in your POR:

- ✓ The *function* of the product is clearly articulated.
- ✓ To avoid feature creep, a *baseline*, minimally acceptable feature set has been identified with other wished for features added to *stretch goals*.
- ✓ A *sketch* has been completed of the anticipated product and a block diagram of the major functional blocks.
- ✓ A rough schematic is provided with the most important components identified.
- ✓ Some metric for what it means to *work* is documented.
- ✓ All *major, noncommodity parts have been identified* with pricing, lead time, and datasheets.
- ✓ Some *cost budget* has been identified and a rough *estimate* completed to verify the costs will be within budget.
- ✓ A *rough power budget* has been established with an idea of the worst-case power consumption of each component and how much voltage and current are required by the power delivery components.

- ✓ A *bring-up plan* has been documented, listing some of the features to be incorporated to make bring-up and troubleshoot easier.
- ✓ A *final test plan* is documented that identifies what will be measured, characterized, and tested, in order to verify performance after it is established the product works.
- ✓ A *risk assessment* has been done resulting in a documented list of the major problems that could go wrong with some indication of how to avoid these problems identified.
- ✓ *Risk reduction* steps have been planned out, such as building some circuit elements in a solderless breadboard, performing SPICE or other simulations, purchasing evaluation boards of some components, or a reduced-function version of the higher risk elements of the product as a proof of concept (POC) version.
- ✓ If you plan to build a solderless breadboard version of a part or all of your final product, you have made arrangements to turn any SMT parts into through-hole parts with 100 mil centers.
- ✓ A *schedule* of what must be completed by what date has been written and mapped into your calendar.
- ✓ If substantial *software development* is required, a parallel software development plan should be created that could use an early prototype version of the product or a modularized plug-and-play version just to develop and test the software.
- ✓ Near the completion of the POR, a *preliminary design review* (PDR) has been completed with another pair of eyes and with management before significant money has been spent or committed.

7.2 Step 2: Create the BOM

The *preliminary* bill of materials (BOM) lists all the noncommodity parts, such as the microcontroller, sensors, actuators, any wireless components, and any special power supply considerations and

7.2 Step 2: Create the BOM

display devices your product will use for which there is even the slightest concern about procuring in the time frame you have.

In this step:

- ✓ The BOM has been created listing the major components, the vendor, and a link to Digikey or other vendors, with lead time and price information.
- ✓ If the assembly will be done by an external fab or assembly shop, the details of what parts they can assemble, or how you will get parts to them, has been established.
- ✓ If assembly is to be done by an external shop that will only use their parts, you have access to their approved parts and can select all or most of the parts you need from their inventory.
- ✓ The cost and delivery time of each noncommodity part has been checked and is acceptable.
- ✓ If you plan to use a BGA or QFN component, you have decided how to get these parts assembled to your board. If you plan on manually assembly them, this is a risk site and a risk mitigation plan should be in place.
- ✓ The specific parts that are still to be added to the BOM, such as commodity parts, resistors and capacitors, have been articulated and put on the to-do list.
- ✓ The datasheet for each major component has been pulled up and reviewed, highlighting any concerns.
- ✓ Any red flags or concerns for any component has been raised and identified and added to the POR. In some cases, important design decisions should be answered by checking the datasheet, the vendor's website, Google, or by planning to build a solderless breadboard or other prototype.
- ✓ You have the symbols and footprints of every part you intend to use in your EDA tool PCB library.
- ✓ Any part not currently in the library might be found from external sources such as snapEDA, Octopart, or the Altium vault.

- ✓ Any symbols or footprints you need that you cannot find already available, you will have to create yourself.

7.3 Step 3: Complete the Schematic

Schematic capture is the process of translating the block diagram sketch of the function of the widget, or the rough schematic, into the final and complete electrical description of the widget. If it is not in the schematic, it is not in the final part.

The schematic is a graphical description of each component used, each of its pins, and how all the pins are connected together to create the net list. The net list is the final database of the connectivity of your board.

In order to add a part to your schematic, you must have the symbol for the part and its footprint in your library.

- ✓ All the major parts have been placed on the schematic page(s).
- ✓ All the parts have been connected up correctly.
- ✓ Net names with readable labels have been created, where appropriate.
- ✓ The obvious problems have been checked for and they are not present.
- ✓ Every single part is included in the design tool's library with its symbol and footprint.
- ✓ The footprint of all the major parts have been compared with the actual board footprints on the library, simply by printing a 1:1 footprint and comparing it to the part of the mechanical drawing of the part.
- ✓ The schematic pages are neat and organized in a way that anyone can follow with a consistent flow for power and signals.
- ✓ Net classes for specific types of nets based on special layout considerations are identified.

- ✓ Specific concerns or directions for layout have been noted on the schematic page.
- ✓ Any remaining concerns or risk sites about how a part should be hooked up or used is identified and listed in the updated POR.
- ✓ The schematic has been checked against the POR and the functioning of the product to verify that if it works it will meet the performance goals.
- ✓ The POR is reviewed to verify the schematic is consistent with the POR.

7.4 Step 4: Complete the Layout, Order the Parts

The layout is the physical implementation of the board. It is the directions the fab shop needs to manufacture the bare printed circuit board that will meet all the DFX criteria to give you the best chance of success.

In this step the following tasks will be accomplished:

- ✓ A fab shop has been identified and their design rules noted. This includes the stackup they manufacture at the lowest price.
- ✓ The design rules for the fab shop have been set up in the constraint manager of your EDA PCB design tool.
- ✓ The board outline has been defined and implemented, usually on the keepout layer, 20 mils wide.
- ✓ The layers in your stackup have been assigned, with the preference to keep all layers as signal type and implement copper pour as the planes.
- ✓ The parts that need to be corner or edge pieces have been identified and placed where they should go, such as connectors, antennas, and test points.

- ✓ Critical parts have been identified and placed on the board where appropriate, such as LEDs, switches, and interface chips.
- ✓ Parts with many routing connections have been placed and translated and rotated around to try to reduce the crossovers in the ghost traces.
- ✓ All decoupling capacitors are in close proximity to their ICs with short, wide paths between the capacitor pads and the IC pads for both power and gnd connections.
- ✓ The most noise-sensitive power pins have ferrite filters and decoupling capacitors on their power pins.
- ✓ Ground vias have been placed in close proximity to the pad to which they connect.
- ✓ All the signal lines connected by ghost nets have been routed with 6 mil wide traces.
- ✓ The number of cross-unders is minimized by clever routing topologies.
- ✓ Cross-unders and the gaps they create in the ground plane are short.
- ✓ Any unavoidable, long gaps in the return plane have return straps spanning over the gaps.
- ✓ All the power paths have been routed with 20 mil wide traces.
- ✓ No large-area copper pours are used on signal layers.
- ✓ All VRMs have their appropriate filter capacitors in close proximity.
- ✓ All components have silk screen labels with polarity as needed.
- ✓ All indicator LEDs, test points, and switches are labeled with intelligent names any user can interpret without looking it up in the manual.
- ✓ All labels needed for end users are placed and readable.

- ✓ Your board has silk screen with your name, the board name or layout file, rev or date code, and other identifiers or logos.
- ✓ At the completion of the layout, a CDR (critical design review) has been completed with another pair of eyes, and the Gerbers reviewed.
- ✓ The Gerbers and assembly files have been sent to the fab shop for DFM verification and the quote is acceptable.
- ✓ Be mindful of any features you are specifying that add cost to your board.
- ✓ Unless you have a strong compelling reason otherwise, use the lowest-cost, default conditions for your board features so it does not add any additional cost.
- ✓ If the fab vendor is doing the assembly, the BOM with the appropriate part identification is created and the pick-and-place file is created in the correct format.
- ✓ The order has been placed for the fab and assembly and the delivery date is acceptable.
- ✓ All the parts in the BOM have been ordered and the delivery dates are acceptable.
- ✓ Any concerns on delivery or cost issues have been raised, with a plan B or plan C in the works as back-up.

7.5 Steps 5 and 6: Assembly and Bring-Up

In these steps, the components are added to the bare board, turning the printed circuit board (PCB) into a printed circuit assembly (PCA).

In the prototype phase, the first time the board has been designed and built, performing the bring-up and test should be done with the board assembled in stages.

Before you add a new functional block, you can test and confirm correct operation of what is already assembled and then move on to the next components. This is only practical if you are doing the assembly yourself. This is the most efficient way of isolating the potential root cause of a problem.

If your board is assembled by an outside shop, you obviously do not have this opportunity. As an alternative, it is possible to add switches in the power distribution paths or signal distribution paths to specific circuit blocks so they can be selectively powered on or off. This allows some element of isolating circuit blocks. Just be aware of the potential problem of an input voltage appearing at an I/O to a device that is powered off.

The tasks to be completed in this phase include:

- ✓ The bare board has gone through an initial visual inspection and no issues found.
- ✓ Pictures have been taken of the bare PCB and the assembled board.
- ✓ The bare board has passed an opens/shorts test, especially for the power and ground.
- ✓ All the parts have been received.
- ✓ The assembly method has been determined.
- ✓ The power generator parts have been added and successfully tested.
- ✓ The initial parts in the signal chain have been added and successfully tested as they are assembled.
- ✓ The solder joints of all leaded ICs have been inspected under a microscope for good solder joint quality.
- ✓ All isolation switches have been opened to isolate circuits and each circuit, one by one, has been turned on to verify performance.
- ✓ The board is functioning or working as expected based on the functional spec and definition of "working" in the POR.
- ✓ The noise levels on power rails and signals have been measured.

- ✓ Where possible, switching noise on quiet-low and quiet-high pins have been measured.
- ✓ All measurements have been completed, with important figures of merit extracted.
- ✓ All consistency tests have been conducted to support your analysis.
- ✓ All stress tests have been completed.
- ✓ Pictures and screen captures of scope traces or other images have been collected for the final report.
- ✓ When you encountered problems, you applied troubleshooting techniques to find the root cause and fix the problems.
- ✓ You have added to your list of possible errors all the ones you encountered in this design so that you will watch for and not make these mistakes again.

7.6 Step 7: Documentation

The documentation step is how you rachet up the learning curve. This is where you capture and document what you learned in this product design cycle so that you can leverage new habits or potential risk reduction actions for the next design.

You should be able to take your plan of record and add to it all the things you wish you had done before you started each step of the project so that you would have avoided a problem, a risk, or made your job easier. Whenever you think, "easier," think "lower risk."

The ideally designed product would be one in which you can throw all the parts you want to use into a paper bag, shake it up, and have the fully assembled, perfectly working widget fall out of the bag. The closer your product development path is to this ideal process, the easier it is, and potentially the lower the risk.

From each product development cycle, you should be asking, "what can I do next time to make my next product design cycle closer to this ideal?"

In the documentation phase, you are reviewing what you did in this project, how well did the final path match your plan, and what you can do differently next time to achieve a better performance-cost trade-off at lower risk and shorter schedule.

The documentation for a project generally is of two formats: a white paper you write for yourself and a formal document you write for broader distribution.

The formal document may be a publication in a journal or conference, it may be for an internal memo distributed inside your company or to a customer it may be as a report for a class. These documents will have their own criteria of what details should be included, how long they are, and who the audience will be.

Regardless of the required documentation for external consumption, you should get in the habit of writing up a white paper for yourself that is a core dump of everything you did and learned in this project.

This white paper serves two purposes. The process of writing it will help you think through the details and force you to reflect on what you did right and what you did wrong. What new problems did you encounter and fix that should be added to your list of errors not to make next time? Take the time to think over the entire design process from start to finish.

The second purpose of your white paper is to provide a record you can refer to a week or a month or a year or even five years from now to review the details of what you did and what you learned from the experience. You may wish to reuse some parts of the design or some components and your white paper can use a manual on how to use the sections in your next project.

The more details you add in your white paper, the more you will benefit from the process and the content in the future.

The following are some questions to stimulate your thinking of what you could include in your white paper.

- ✓ How well does the widget you built match your expectations?
- ✓ Did you learn anything from the noise measurements?
- ✓ If you could do this project over again, what would you do differently?
- ✓ Are there features you should have removed to increase your chance of success without decreasing the value much?
- ✓ Are there features you could have added from your stretch goals without sacrificing your chance for success?
- ✓ What could have been added to the schematic to reduce the risk?
- ✓ What could have been added to the silk screen to reduce risk?
- ✓ Are there components you used that you should never use again?
- ✓ Have you updated your list of errors not to make next time?
- ✓ Have you updated your best design practices list?
- ✓ What is the documented performance of your widget?
- ✓ How well does it meet your initial performance expectations?
- ✓ What is the worst-case noise compared to what can be tolerated (the margin)?
- ✓ Are there any special directions on how to use it or what to watch out for?

7.7 Practice Questions

1. What are the seven steps in any board project?
2. What is the main purpose of the POR?

3. Why is it important to hand-sketch the schematic?

4. What is an advantage to assembling the parts to the board manually in early prototype designs?

5. What is a general strategy for the bring-up phase?

6. What are two features to consider adding to make it easier to bring up a board?

7. What are the most important components to include in the preliminary BOM?

8. What is a purpose of the documentation phase?

9. What value is a plan B, plan C, or plan D?

10. What is the downside of designing your project serially?

11. What is the ideally designed product?

Chapter 8
Step 1, POR: Risk Mitigation

The path from a BoN to a working widget is like a chain across a deep canyon. To make the complete path, every link must be in place.

If you are going to walk across the canyon on the chain, you do not want to check that each link is in place *right before you come to it*. You want to check each link *at the beginning*, before you begin the journey, and before you step foot on the first link.

8.1 Visualize the Entire Project Before You Begin

Do not design your board serially, thinking about the next task only after you have completed the previous one. This will guarantee a longer schedule than you can probably afford and the possibility of encountering a roadblock halfway through your project.

If you selected an NXP i.MX RT1060 microcontroller for your product and have designed the schematic and the board and then decide to order the part, you might find the delivery time is 8 weeks. If your project cannot afford an 8-week delivery delay, the technical term for this situation is "you're screwed." If you had known it was an 8-week lead time four weeks ago, you could have either placed the order early, or planned for a different microcontroller.

During the POR, you visualize the ENTIRE project and all the potential problems you expect to see at each step before you begin the project. This allows you the time to pay attention to the potential show-stoppers early enough to do something about them, or to replan the project to detour around them.

This is what *risk assessment and mitigation* is all about. Before you get far into the design of your product, you want to think through all the possible problems and design them out or establish a back-up plan as early in the planning phase as possible. This is what designing like Ralphie's mom and not like a Colorado Bro is all about.

You cannot just assume the product will work perfectly after you design it, do the layout, assemble the parts, and power it on. In the real world there are potential problems that can sneak in at every step along the way. You will need to identify the problems and design them out at each step along the way.

If you have never created a schematic or completed a board layout, it is hard to know problems to anticipate and design out. This is where the checklists in this book are valuable.

This is also why it is so important to follow the guiding principle of prototype development: *fail early and fail often*. This will give you the opportunity to accelerate up the learning curve.

Consider building smaller, less ambitious circuit boards early in the design cycle to make mistakes earlier when they have less impact on the final product.

Think about worst-case scenarios at each step. Plan on designing for:

- Design for connectivity (DFC)- connectivity
- DFP (performance)- DC
- DFP- signal routing
- DFP- power routing
- DFP- EMC
- DFM- manufacturing fab
- DFA- assembly
- DFQ- quality
- DFB- bring-up, debug, and troubleshoot
- DFT- test
- DFU- user experience
- DFR- reliability, long life

8.2 Avoid Feature Creep

In the early phase of brainstorming product features, it is easy to get carried away with adding more features. If everything were free, why not add more features? We call this process of adding more features during the product concept forming phase, without thinking about their cost, *feature creep*.

Wouldn't it be cool if there were flashing red and green lights? Wouldn't it be great if there were wheels and the board were to roll around the floor avoiding obstacles? Maybe there should be a large 4k resolution screen that pops up on demand? Maybe there could be a touch screen the user interacts with, maybe…

But features are not free. They may add a small amount of dollar cost, but more importantly, each feature adds a risk cost, and may add a time cost. Avoid the temptation of trying to do too much and increasing the risk of failure.

When you see all the steps required to complete a successful product, you may realize you do not have time to implement all the features you want, or are not willing to take on the risk all the features would create.

If you must accept feature creep in your design, partition the new features into different generations of the product.

To increase your chance of success in getting the first rev out, think about what the minimum acceptable baseline feature set may be and focus on this. Then you can partition other features in the next few product revs or as stretch goals if time permits.

By thinking of the future evolution of your product, you can anticipate additional features and design the hooks in your product to allow expansion in the next generation.

8.3 Estimate Everything You Can

The way to anticipate the performance and power requirements of your product is by putting in the numbers at every opportunity. This step should be started as early in the design phase as possible and continued throughout the design cycle as often as practical.

Based on your circuit and worst-case use, estimate the current draw of each component and the total power dissipation of the board. You should get a copy of the datasheet for all important parts so you know what voltage and current requirements to expect.

Use simple approximations to get a rough idea. Is the current draw from all your components 1 A or 0.1 A? The power supply you need may be very different depending on the max current requirement or the average current requirement. If battery powered, how long will you expect the battery to last?

What about the power consumption of each part compared to the typical power dissipation rating for each package?

What is the pole frequency of each filter in the power supply and the potential noise spectrum of the sources?

Will your board require transmission line, controlled impedance interconnects? If the longest interconnects in inches are longer than the signal rise time in nsec it doesn't mean you will need controlled impedance interconnects, it just means it will be important to do at least a simple simulation that includes a driver model.

Sometimes you will not have all the information you need to make these estimates. In this case, you may have to make a rough guess, and update your guess as you learn more. Always keep in mind:

> *Sometimes and OK answer NOW! Is better than a good answer late.*

If your rough estimate suggests you are 10x outside the safe zone, it is safer to take the time now to get a better estimate to decide to design around this component than wait until you know for certain you have a problem.

If you are building an array of 20 flashing LEDs controlled by the output of a microcontroller, you should worry about the current draw of each LED and how you could supply potentially 400 mA or even 1000 mA of current to the LEDs. The typical total current you can source from a microcontroller is less than 200 mA.

8.4 Preliminary BOM: Critical Components

Sketch the functional diagram as much as you can, even as just a block diagram. This will help you think through the power flow and the signal and information flow. The better you can conceptualize what each circuit block does, the better you can decide if the design will meet your needs.

Based on the block diagram, sketch the schematic, even if just on the back of a napkin (metaphorically). It is not critical to have all the details, but at least the major components and their connections of signal and power flow.

Identify the most important parts you will need. Find them on digikey.com or whichever source you will use to purchase them. To make purchasing and tracking orders easier, and less risky, try to find all of your parts from the same distributor, like www.digikey.com.

Check delivery time for every critical part. If the lead time is 6 weeks, maybe find another vendor or consider a different part.

Find the datasheets, and explore their features and any specific requirements — software, drivers, etc. Read the datasheets to learn the correct way of using the components and what performance you can anticipate.

8.5 Risk Assessment

In this first phase of the project, we look at *risk assessment* and *risk mitigation*. The combination of these two is *risk management*.

This is as big a part of the design process as is the actual design. There are many more ways of screwing up a board and getting it wrong than doing it right.

You should get in the habit of thinking about the design process like Ralphie's mom, rather than like a Colorado Bro. Get in the habit of thinking of all the worst-case scenarios — everything that could possibly go wrong, before it happens.

The first step in risk management is identifying all the potential problems. Step back and ask yourself, what are all the possible things I can think of that could go wrong?

If you are assembling your board yourself, do you really want to use 0402 parts? You may think it is an act of macho heroism ("here, hold my beer") but in reality, it is a huge risk site. Why not avoid this risk site and use 1206 parts which have a much lower risk of introducing assembly problems? You should have a strong compelling reason to voluntarily take on more risk than absolutely necessary.

Is it possible the footprint for a component in your library is wrong?

Is an enable pin supposed to be pulled HIGH or LOW?

Should an I/O rail be powered by 5 V or 3.3. V?

Should there be a pull-up resistor on the I2C bus?

Are the RX and TX pins connected correctly?

It is hard thinking of worst-case scenarios. This is why you should use every opportunity to practice. Get in the habit of looking for hidden potential failure modes and play out possible plan B, plan C, and plan D contingency strategies.

This is exactly what is done in war gaming exercises. It is about playing out possible scenarios of, "if the enemy does *this*, we will do *this*, and the reaction will be *that*, so we need to be ready with *this*."

Risk management in product design is a lot like playing chess. You need to think as many moves ahead as you can. As your opponent makes a move slightly different than what you anticipated, you adjust your strategy. Your opponent in product design is the real world.

> *While we might think we are wonderful designers and builders, and every project we work on will be designed and executed perfectly to work the first time, few of us in practice can attain these heroic levels of perfection.*

We all make mistakes. No matter how hard we try, we will continue to make mistakes. What we can hope for is to make fewer mistakes in the next design. The better we can anticipate a problem, the better we will be able to implement a plan to not encounter this problem in the current design.

Risk management is a two-step process:

1. Do a *risk assessment* to identify all the possible things that could go wrong.

2. Develop a *risk mitigation* plan to design these potential problems out of the product and its implementation or a Plan B, C, D, or E in case a potential problem does arise.

A valuable exercise at the very beginning of every project is to sit down and make a list of all the possible problems you can think of that could arise. This is where thinking like Ralphie's mom is such a valuable skill.

Then, you can access how likely each risk is: likely, possible, unlikely, black swan (very rare event, requiring multiple simultaneous factors, like the perfect storm or a rogue wave), or never.

> *If you are called a worry wart for raising your concerns, wear it with a badge of honor, that you are doing exactly what you are supposed to do. But also try to offer a risk mitigation solution for each concern.*

8.6 Risk Mitigation: Tented Vias

Rarely is a solution free. At every step you must evaluate how much it is worth paying to mitigate each risk. What is the cost vs the likelihood and the impact of the problem arising and the cost of fixing the problem after the fact? Is it worth paying extra for the insurance? You hope you will not need it, but if there is a problem, the small investment in a backup may be worth it.

You may purchase LEDs from an off-shore vendor with a stated delivery time of 1 week. But the risk they will be delayed is high. Maybe it is worth purchasing a limited number of more expensive LEDs from a local vendor, enough to build one or two boards, just in case your main vendor does not come through.

Sometimes it is not a question of getting in trouble, but what sort of trouble you want to get into. Is it better to be under budget but late on delivery, or over budget and deliver a working product on time? With good planning, we can try to be under budget, on time, with a working product.

If a fix to a potential problem is free, it's always worthwhile to do it. These become habits. Follow the best design practices, the habits, to reduce the risk of potential problems as much as possible.

> *Becoming a successful board designer is to become a master of Murphy's law, which says, "if something can go wrong, it will go wrong." Of course, the corollary to Murphy's law is, "Murphy was an optimist."*

If you become a master of Murphy's law, you will begin to anticipate all the possible problems that could go wrong and plan a solution to avoid the potential problem to prevent it, or a backup plan B or C or D to fix the problem if it does arise.

For example, when a via is placed close to a soldered pad, there is a chance that when soldered manually, some of the solder might short over to the exposed capture pad of the via. This is a problem that can be anticipated, and a mitigation plan put in place. An example of the

8.6 Risk Mitigation: Tented Vias

placement of vias near a soldered pad that might cause a problem is shown in **Figure 8.1**.

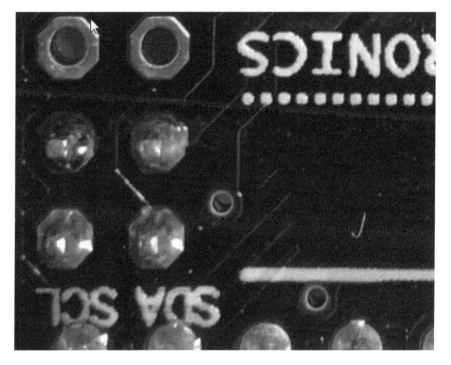

Figure 8.1 Closeup view of a circuit board with vias in close proximity to pins that are soldered. There is the slight danger of solder bridging to these exposed vias.

One mitigation step for this specific problem is to move the via far enough away to not pose a shorting problem. Another mitigation for this problem is to use a tented via.

When vias are manufactured in a circuit board, a capture pad is used at the top and bottom of the vias. This compensates for any registration misalignment tolerances between the drill position and the trace connecting to the via.

After the board is drilled, the top and bottom surfaces are patterned. The final step is to coat soldermask over the surface exposing those regions that will receive the final surface finish, such as HASL.

If the capture pad of a via has an opening in the soldermask, it will be coated by the final HASL or other surface finish and be exposed on the final board. This will leave the via capture pads exposed on the top and bottom surfaces.

The advantage of this approach is that every exposed via is a potential test point or connection point for an engineering change wire during the bring-up, debug, and troubleshoot phases.

The downside of leaving a via's capture pad exposed to the HASL surface finish is that it might potentially short to a component pad during manual soldering if it is too close.

If this is a concern, a specific via can be selected as tented. This means that the soldermask covers and tents the via hole. The capture pad is not exposed to the HASL surface and remains an insulated surface. An example of boards with untented vias and tented vias is shown in **Figure 8.2**.

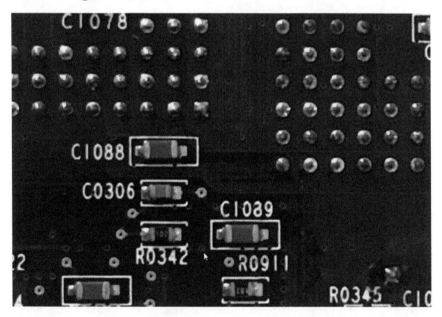

Figure 8.2 Closeup view of a circuit board with some tented and some untented vias. In the upper region are arrays of connector pins. Note the faint tented vias near these soldered pins. Tenting them reduces the risk of unintentional shorting when soldering the pins.

8.6 Risk Mitigation: Tented Vias

When there is the risk of potential shorting, a tented via can be selected for those vias at risk.

The default configuration for your board can be untented, and only those vias with a potential for shorting can be selected as tented. **Figure 8.3** shows an example of how to select a tented or untented via when placing a via in Altium Designer, for example.

Figure 8.3 An example of selecting a tented or untented via in the properties of a via in the Altium Designer layer out tool.

There is rarely a cost difference between selecting a tented or untented via since it is just a specific feature in the soldermask layer.

It is important to pick one default and stick with it. Should you use tented or untented as your default condition? This is a personal preference. When doing prototypes each untented via is a potential test point and engineering change wire terminal. This is a benefit.

When doing a production board, these features are unnecessary and tented vias might be lower risk.

8.7 Risk Mitigation: Qualified Parts

Whenever selecting a new part that you have never used, there is also the risk the part or its footprint does not behave as expected. The consequence is that you don't realize the problem with this part until the board comes back from fab and is tested. The component, and the board, have to be replaced, or the board redesigned and the schedule slips by as much as 2 weeks and there is the extra cost of a board respin.

One way of mitigating this risk is to only use qualified parts. As part of a risk reduction process, you should test each part before it is committed to use on a board. This can be by purchasing a few samples of each part, or even building a small eval board as part of the POC phase.

An example of an evaluation board for a variety of resistors, capacitors, and LEDs is shown in **Figure 8.4**. In this board, the footprint and polarity are tested for each part and a DMM can be used to measure the value of each part.

8.7 Risk Mitigation: Qualified Parts

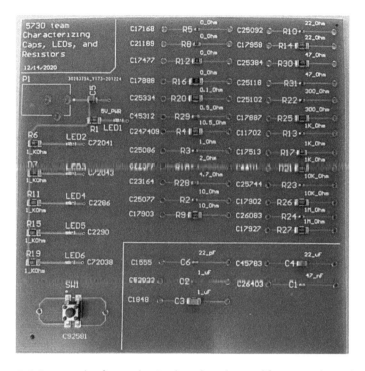

Figure 8.4 An example of an evaluation board used to qualify commonly used parts such as resistors, capacitors, and LEDs. Board designed by Abraham Payan.

An important best design practice is to reuse a part you are familiar with from a past design in your current design. In a large company, this called a "qualified" part. It has less risk since you know what to expect.

If you are on a small team, each new part you use in your projects and have confidence in could be added to your favorites list of qualified parts.

The fewer new, unqualified, unique parts you use in your design, the lower the risk for an unforeseen issue with the component that would arise in your new design.

In the POR, we will also apply this idea of identifying potential risk sites to issues related to attaching parts to the circuit board with solder. Take a moment to think about some of the potential risks of what could go wrong when it comes to soldering parts to a board. Here is my list:

- ✓ The part and its footprint do not match.
- ✓ One part is too close to another on the board and interferes with the soldering of the smaller part.
- ✓ You selected or purchased the wrong part.
- ✓ The part is too small and you cannot manipulate it to place it on the tiny pads.
- ✓ The part is too big and will not fit on the pads.
- ✓ The solder joint you made is bad or a lead is lifted off the board.
- ✓ The lead pitch is too small and you cannot solder to individual leads without shorting to other leads.
- ✓ The leads are under the package and you cannot align the part to the board or solder it to the pads.
- ✓ Polarized parts such as LED are assembled in the wrong orientation.
- ✓ A via is too close to the pad being soldered and gets shorted to the component.

Using only qualified parts will reduce these risks, but may limit the performance or features of your board, or increase the schedule due to delay in qualifying parts.

What are some of the actions you could take to reduce the risk of each of these potential problems ever happening, or fix them once the parts and board are assembled?

8.8 Practice Questions

1. What is feature creep and what is one possible consequence of it?

8.8 Practice Questions

2. What are three possible risks to watch out for?
3. Give two examples of risk mitigation steps.
4. What does it mean to design more like Ralphie's mom?
5. What parts should be included in the preliminary BOM?
6. What is a qualified part and why is this important?
7. What is a plan A, B, and C?
8. What is wrong with designing your product serially?
9. How do you become a master of Murphy's law?
10. What is risk mitigation?

Chapter 9
Risk Reduction: Datasheets, Reverse Engineering, and Component Selection

Almost every prototype design you build will be custom in some way because of your specific set of conditions.

If you are following a reference design or a circuit design you find on the internet, maybe you do not have the 13.7k ohm resistors the reference design used. Maybe the specific 555 timer part used in the internet circuit is not defined or is different from the one you have available.

This means that even if you try to duplicate a reference design, you may not be able to follow it exactly. You will have to perform your own engineering analysis of the circuit.

9.1 Take Responsibility for Your Design

Ultimately, every prototype you build you have to make your own and take personal responsibility for it. There is no substitute for understanding the circuit design and the motivations for the circuit design and the component selections and their values. The better you understand the behavior of the circuit and the required performance of your circuit, the better you will be able to make design decision trade-offs.

Not all pin compatible components are the same. Even though two ICs may be labeled as 555 timer chips, they may have very different characteristics, such as current draw, output impedance, and rise time.

When you use a component in a design, you don't want any surprises. You want to learn as much as you can about the part before it comes time to assemble it onto the circuit board.

There are a number of steps you can take to reduce the risk of unexpected surprises when it comes time for the final test.

When you read a datasheet or use a reference design, the most important skill you can apply is to read the documents *critically*. Take responsibility for your interpretation of the document and what you take away as valuable information.

Just because it has been posted somewhere does not mean that it is correct or even if it applies to your application. Critical reading is a must before you accept what you read in any documentation.

9.2 Reducing the Risk of a Design Problem

There are six activities you can do to reduce the risk your design will not work. Be aware of the schedule impact of these activities. The benefit of risk reduction at the cost of time and extra expense needs to be considered.

It may be a better trade-off to extend the schedule or increase the budget to reduce the risk. This is sometimes referred to as "buying insurance." We can often pay extra in time and money to reduce risk.

This decision needs to be factored into your product development cycle and budgeted as early in the design process as possible.

The following six actions will reduce the risk of an incorrect design or selecting the wrong part for your application:

1. Review the datasheets, application notes, and reference platform designs of your critical parts (very important!).

2. Research the parts and circuits as much as you can online, but realize that just because something is posted online does not mean it is correct!

3. Simulate the circuit with a SPICE or other simulation tool (assuming you have the correct models for all the components).

4. Build a solderless breadboard prototype (assuming the interconnects are transparent in your application).

5. Build or purchase an evaluation board with your major components, power it up, and test its performance, (if you can fit this into your schedule and budget).

6. Purchase a modularized system that you can connect together to hack or modify a comparable system with the appropriate software as you want to build.

The most inefficient process, which takes the longest time and has the highest risk of not working the first time, is to build your product with a suspect component or circuit and plan on testing it after it is built. If you find something does not work as expected, you have to find the root cause, change the part, or try to repair the circuit or change the board fab or select and order other parts. This will take a lot of time. This process is called "build it and test it." It is notoriously inefficient.

This will happen naturally in the product development cycle. If you want to increase your chance of success for this phase, don't rely on "built it and test it" multiple times to achieve a successful product.

9.3 Understand Your Circuit

When designing a new circuit, there will always be some questions about the right or wrong way of connecting a device to achieve the desired performance. There is no substitute to thoroughly understanding your circuit.

Always take responsibility for your own design. Just because a reference design does it one way does not mean this is the right way.

Just because you read a forum post that suggests one option does not mean this is the best design practice, or that it will even work.

For example, you will inevitably see recommended on many reference designs to use three capacitor values, a 0.1 uF, a 1 uF, and a 10 uF capacitor. This is based on 50-year-old assumptions about using through-hole capacitors that do not apply to surface-mount capacitors. Instead, you should use a single, large-value, low mounting inductance capacitor. If you have the option of using three capacitors, you should select the same large-value capacitors. The reasoning behind this is described in this article, and later in this book.

If you rely on a Google search and the information posted on forums, you run the risk that the post could be wrong.

For example, in a simple 555 astable vibration circuit, should pin 5, the CONT pin, be left open or connected to a capacitor, and what value capacitor should be used? **Figure 9.1** is an example of three different circuit recommendations taken off the web for an astable vibrator using a 555, showing three different recommendations for the connection to pin 5, the CONT pin. Which one should you follow?

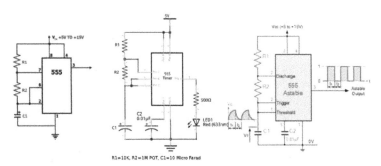

Figure 9.1 Examples of three different 555 astable circuits taken off the web.

You should never blindly follow any recommendation you get from the web or a reference design, but take responsibility for your own decisions. The more you understand your circuit, the better you will be able to make correct design decisions.

9.3 Understand Your Circuit

In the case of the 555 timer, insight into what should be connected to pin 5 can be gained by looking at the internal circuit of the 555, taken from the datasheet and shown in **Figure 9.2**.

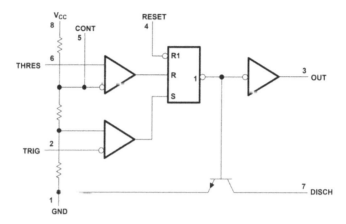

Figure 9.2 Schematic of the inside of a 555 showing where each pin connects, taken from the datasheet.

We see clearly that pin 5 is connected to a reference voltage pin equal to about 2/3 Vcc. This pin is used to define the switching threshold that the comparator will use. When the THRES pin goes above 2/3 Vcc, the comparator will turn on and trigger the R input to the flipflop.

From this circuit, we see the purpose of the CONT pin is to bring out the comparator reference voltage. Based on this circuit, what should be done with pin 5?

In principle, pin 5 can be left open. The 555 astable vibrator will work just fine with pin 5 left open. This is why some designs show no connection.

However, in practice, there may be a little noise on pin 5 picked up from the internal switching of the 555. To reduce this very slight noise, a filter capacitor can be placed on pin 5. This is why some schematics show a 10 nF capacitor. What is special about a 10 nF capacitor? The value determines the pole frequency, in addition to the series R that is part of the voltage divider. A higher-value capacitor means a lower pole frequency and filtering more noise.

But, a higher value capacitor will also have a longer charging time constant to turn the threshold voltage on and reach a stable oscillation frequency after power on.

If the transient turn-on time is not important, a larger value may work just fine. Not knowing the series resistor values means we have no way of knowing what the transient charging time might be. This is a case where unless there is a strong compelling reason, following the recommendation of 10 nF capacitor or an approximate value might be acceptable.

9.4 Read Datasheets Critically

One of the limitations of datasheets is that they are sometimes written very conservatively. In order to have a higher yield from production, the specifications are written with broad parameters values so more parts meet the spec.

Sometimes these specs are so broad as to be worthless in helping you decide if this part will work in your application. The rise and fall time specified usually varies by a factor of 2 or 3. If you are looking for a part with a 10 nsec rise time and the range is 10 to 40 nsec, will your part really have a fast enough rise time?

Even worse, if you want the longest rise time to reduce switching noise, should you hope for a part with 40 nsec, when they may in fact all be closer to 10 nsec?

Sometimes the only way to know how a typical part will actually behave is to purchase some and test them. This is where either a solderless breadboard or an eval board or a simple prototype can be an important risk reduction.

9.5 Build Simple Evaluation Prototypes

Sometimes, the parameter you are most interested in is not even listed in the datasheet, such as an output Thevenin resistance. The more common spec is the output source or sink current. Rarely are the conditions under which this term is defined clearly stated.

Is it the maximum current when the output is connected to Vcc or to ground? Is it the current at which the output voltage drops by 10% or is it the maximum output current based on the power consumption of the part and its assumed thermal management? For a specific part, any of these definitions might apply.

Sometimes the only way to answer this question is to buy a sample part and measure the output Thevenin resistance, the voltage drop at maximum current, or the case temperature at maximum current.

Rarely is the rise or fall time of a driver listed in a data sheet. The propagation delay, which is the time delay between an input signal to a corresponding output signal, is often listed. While this is related to the rise and fall time, it is not the same.

For these reasons, it is sometimes important to purchase some test units and build a simple evaluation circuit to measure the performance.

For example, not all 555 timers are the same. The rising edge of a NE555 timer chip is about 200 nsec. The rising edge of an LMC555 chip is 50 nsec. One is rated at operating up to 100 kHz and the other up to 1 MHz. They are both 555 timer chips. It's only by carefully reading the datasheet that we can determine which chip is suitable for an application and even then, sometimes only by measuring an actual unit.

9.6 Reverse Engineer Components

There are sometimes important design parameters that are not easily available in a datasheet. For example, the output impedance of a power supply, or the rise time of an I/O.

When modeling a DC source, there are a number of important parameters we might care about:

- ✓ The Thevenin voltage

- ✓ The Thevenin resistance
- ✓ Any current limiting element
- ✓ The switching frequency
- ✓ The unloaded ripple noise
- ✓ The loaded ripple noise at some DC current
- ✓ Any high-frequency noise
- ✓ The transient response to a step current load

It is difficult to get some of this information for all voltage sources. Sometimes the only way is to build an eval board and measure it. We match the measurements to a simple model and extract the values of the parameters that describe its properties. This is reverse engineering.

In Chapter 6, we described how to measure the Thevenin voltage and resistance of any voltage source. This is reverse engineering the important figures of merit of the voltage source.

The propagation delay of a 74AC14 hex inverter is listed in the datasheet as 1.5 nsec to 10 nsec. This is a range of more than 6x. It is way too broad to base any timing critical design. The only way to know what is typical is to build up a small evaluation vehicle and measure it.

The propagation delay of an invert is easily measured by constructing a ring oscillator with an odd number of gates and measuring the self-oscillation frequency. The ring oscillator circuit schematic and the actual ring oscillator circuit board is shown in **Figure 9.3**.

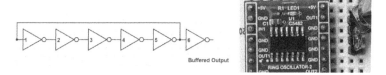

Figure 9.3 Schematic of a ring oscillator and the actual ring oscillator design by Chaithra Suresh.

The self-oscillation period with 5 inverters in series is 2 x 5 x PropDelay. The output signal from the buffer was measured by a high bandwidth scope with a period of 20.7 nsec. This makes the propagation delay 2.07 nsec. While this is within the 1.5 nsec to 10 nsec range of the specification in the data sheet, it is a much better estimate of the propagation delay to expect for this part. At the same time, the rise time of the signal was measured as 1.36 nsec. This measurement is shown in **Figure 9.4**.

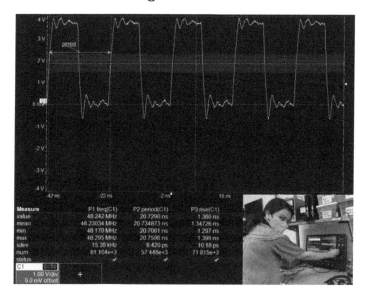

Figure 9.4 The actual scope measurements of the period and rise time of the buffered signal out, measured by Chaithra Suresh, shown in the inset.

When an accurate performance metric of a component or subsystem is needed, sometimes the only way of reducing the risk of the uncertainty is by reverse engineering a parameter from the measurement of a test vehicle.

9.7 Reuse Parts

The uncertainty in the actual performance of a new part you have never used before can be a significant risk site. An important way of reducing this risk is reuse a part you have already used and have confidence in. In other words, only use qualified parts.

In larger companies, every component goes through an extensive qualification process with component characterization, vendor qualification, and a process in place to make sure the vendor does not change a part without approval.

In a smaller group, each engineer can keep a qualification list, usually referred to as a "favorites." Each time you use a new part and are able to verify its footprint, rise time, output impedance, and other important features, it can be added to your favorites and reused with confidence.

If you have multiple new parts you plan to use in a design, it may be an important risk reduction step to build a simple test board and purchase the parts to test them out. Then they can be added to your favorites list.

Use each product you build using your favorite parts to rachet up the learning curve. Leverage every design as much as possible as a learning opportunity to study, understand, and verify the performance of each part so you know what to expect from it.

A second important strategy is to *minimize the number of unique parts* you use in your design. This is sometimes referred to as *scrubbing the BOM*.

Reducing the number of unique parts used in your board will:

- ✓ Reduce the risk of unexpected performance

9.7 Reuse Parts

- ✓ Reduce the possible confusion (risk) of placing the wrong part on the board
- ✓ Reduce the complexity of managing the different parts and tracking their delivery (risk)
- ✓ Reduce the costs by ordering a higher quantity of the same part
- ✓ Reduce the automated assembly cost by requiring fewer reels of parts on the pick-and-place assembly machines.

For example, in selecting decoupling capacitors, use a MLCC capacitor, which has a voltage rating at least 2x the application voltage. This provides some safety margin and keeps the capacitor in the voltage regime where its capacitance is close to its rated value.

Select the smallest body size MLCC capacitor consistent with your assembly method. If you are assembling the board yourself by hand, maybe you will want to use 1206 size parts. If by automated assembly, consider an 0402 size capacitor. Find the largest capacity capacitor in that body size at the right voltage rating and at the lowest acceptable cost.

This generally translates to a 10 uF to 22 uF capacitor. Wherever you plan on a decoupling capacitor, use this part in as many applications as practical.

If you have an application for another capacitor, try to use this same 10 uF capacitor value if at all possible. If your timing circuit calls for a 1 uF capacitor, consider using the same 10 uF capacitor value as used for decoupling.

If the precise RC value is not important, consider using a value of the resistor that you already have available and the 10 uF capacitor already being used for decoupling.

If you control the specifications for your product, balance the trade-offs between component values based on parts you are already using and still stay within an acceptable performance range.

Selecting parts and their values to use in a design based on the trade-off between re-using parts, using parts that are available in a library of acceptable parts, yet still meeting the acceptable performance, is one of the most important skills of any engineer.

The secret for being successful in this trade-off selection is to know your circuit. Just because the reference design suggests some component values or the datasheet offers a recommendation or a circuit you find on the internet has some specific set of values, should not be the reason you have to rigidly adhere to these values. Make your own decisions.

Sometimes there are historical reasons for specific values with assumptions that no longer apply. Sometimes there are reasons based on avoiding a common or sometimes very rare problem that might arise. Investigate why specific values are chosen and make your own decisions based on your customer design.

9.8 Practice Questions

1. What is an example of a problem you can encounter when using a schematic you find on the internet?

2. What does it mean to take responsibility for your own design?

3. If you have never used a specific circuit before, what are three things you could do to increase the chance of having it work the first time?

4. What is an example of a feature of a part you might care about that you would not find in a datasheet?

5. What should you do with pin 5 in a 555 timer?

6. A DMM only measures voltage. Yet, it will display current. How does it do this? What consistency tests could you try to reverse engineer what is going on inside the DMM?

7. A DMM only measures voltage. Yet it will display resistance. How does it do this? What consistency tests could you try to reverse engineer what is going on inside the DMM?

8. When should you pay extra for an evaluation board for a component?

9. What does a ring oscillator tell you?

10. Why should you keep a favorites list of parts and reuse them?

Chapter 10
Risk Reduction: Virtual and Real Prototypes

If your product involves a brand-new circuit or uses a new component for which you do not have past experience, there is a risk it may not work.

- ✓ Maybe your circuit design is wrong.
- ✓ Maybe the part does not perform the way you think it does.

What can you do to minimize the risk of this problem?

There are generally four actions you can take to reduce the risk of unexpected behaviors:

1. Build a *virtual prototype* of your circuit using a circuit simulator like SPICE.

2. Build a *real prototype* using a solderless breadboard.

3. Buy a real prototype in the form of an *evaluation board* or modules that can be integrated together.

4. Design and build a *small custom* evaluation PCB prototype with minimum functionality and complexity to test out the performance of the part.

10.1   Getting Started with Circuit Simulation

A circuit simulation tool will calculate the voltages or currents on any node of a circuit in the time domain (transient simulation) or frequency domain (AC simulation). If you can draw the circuit composed of combinations of ideal component models, ideal sources, and R, L, and C components, a circuit simulator will display the voltages and currents on any node in the time or frequency domains.

The value of performing a SPICE simulation of a circuit is threefold:

1. If you have accurate models for your components, the SPICE simulation will indicate if the circuit performs the way you expect. If you are not familiar with the circuit, the SPICE simulation can act as a virtual prototype to explore the functioning of the circuit.

2. With confidence in the simulation, you can explore the "what if?" questions. What if I don't have a 13.7k resistor, but only a 10k resistor. Will the performance be acceptable? How much flexibility do you have in the selection of components? Use this virtual prototype to balance the trade-off decisions in component value selection.

3. How stable is the design? If component values will vary due to manufacturing variations, or uncertainty in the specs, will my circuit still perform the way I expect it to?

Of course, the important qualifier in using any simulation is the quality of the models used in your circuit. The only elements available are ideal circuit elements that are also included in the version of SPICE you use.

Do you have confidence these ideal circuit elements, and their combinations, predict the behavior of your circuit with acceptable accuracy? It is always worth exploring some level of measurement-simulation correlation either by your own work or searching for published examples to establish some level of confidence in the quality of the simulations, in the best case. This is part of risk reduction.

The most common circuit simulation tool is SPICE (Simulation Program with Integrated Circuit Emphasis). It was developed and first released in 1972 to help predict the behavior of transistor circuits before fabricated on wafer.

It was initially developed at UC Berkeley as open-source code. Versions spun off from and compatible with the original version are

10.1 Getting Started with Circuit Simulation

sometimes referred to as Berkeley SPICE. Since 1972, it has gone through hundreds of modifications and been re-released in open-source and commercial versions.

If you can draw the circuit with ideal elements, SPICE will simulate the predicted waveforms. An example of a circuit and its simulated frequency and time domain performance compared to the real circuit measured with a scope is shown in **Figure 10.1**.

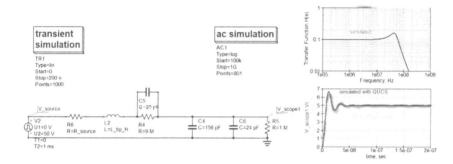

Figure 10.1 An example of a circuit and its simulated transfer function and transient response simulated in QUCS and compared with the actual device's response measured with a scope.

There are many versions of SPICE. Five popular ones are:

- ✓ CircuitLab (https://www.circuitlab.com)
- ✓ LTSPICE (https://www.analog.com/en/design-center/design-tools-and-calculators/ltspice-simulator.html)
- ✓ PSPICE (https://www.ti.com/tool/PSPICE-FOR-TI)
- ✓ QUCS (http://qucs.sourceforge.net/)
- ✓ Keysight ADS (https://www.keysight.com/us/en/products/software/pathwave-design-software/pathwave-advanced-design-system.html)

While CircuitLab, LTSPICE, PSPICE, and QUCS are free tools, they differ in features, performance, and ease of use. They also differ in the variety of models for common IC components available.

Keysight ADS is a commercial tool and is not free but is incredibly powerful.

Regardless of which one you use you should pick one and get familiar with it.

SPICE is to a circuit designer what a calculator is to an accountant. It is a valuable tool but is no substitute for understanding the rules of tax deductions. Every engineer should be comfortable with at least one SPICE tool and have it available to do quick simulations as a starting place for every design.

If the SPICE simulation meets your specifications, it is not a guarantee your design will work. But it will build your confidence that at least the simulation did not find an obvious problem. The more accurate the models you use for the components and the interconnects, the higher your confidence your design will work.

A well supported free version of SPICE was developed by Linear Tech in 2001. It was originally called Linear Tech SPICE, or LTSPICE. After Linear Tech was acquired by Analog Devices Inc (ADI), the name LTSPICE was kept. Under the guidance of ADI, its support has grown and all ADI circuit components are available as LTSPICE models.

LTSPICE is not the only version of SPICE that is free, easy to use, and very capable. I also recommend QUCS. It has a few additional features that LTSPICE does not have, such as native support for S-parameters and built-in approximations for transmission line models. But it lacks a large library of IC models. Its graphical user interface is easier to use than LTSPICE and of publication quality.

Regardless of which version you use, it is important for every engineer to have access to a free version, then get familiar with it and use it on a regular basis.

If you are not familiar with SPICE simulation or specifically with LTSPICE, you can view this series of video tutorials from Analog Devices, starting with this one and working through the 15 videos in the series. They are each about 5 minutes long, so in an hour and a half you can learn all you need to know about getting started with LTSPICE circuit simulation.

Just note when you are viewing the videos from the ADI website, view them from the last in the list first. The latest, more advanced videos are positioned at the top of the list.

Watch this video and I will show you how to set up a simple RLC circuit in the free simulator, Circuit Lab, and run a simulation.

10.2 Practice Safe Simulation

It is easy to get a result from a simulation or a measurement. Sometimes it is as easy as pushing one mouse click and you will see a voltage or current waveform.

But there are many sources of artifacts that can arise which make the results worse than meaningless, misleading. This is the origin of the editorial comment, created specifically to describe the results from simulation, garbage in, garbage out (GIGO). This is illustrated in **Figure 10.2**.

Figure 10.2 An example of the metaphor garbage in, garbage out.

One way of reducing the risk of introducing artifacts in simulations and measurements, and to help identify when you may have made a careless mistake using the tool, is Bogatin's Rule #9. (see, for examples, https://www.signalintegrityjournal.com/blogs/4-eric-bogatin-signal-integrity-journal-technical-editor/post/1539-bogatins-20-rules-for-engineers)

> *Rule #9: Never do a measurement or simulation before you first anticipate what you expect to see. Anticipate before you simulate.*

If you are wrong and the result is not what you expected, do not proceed with the result until you can understand why it is not as expected. Maybe you typed in 17 but meant 71. Maybe channel 1 is not really connected to the signal pin, but to a ground pin. Maybe the scale is incorrect, and you can't see the noise you expect on that scale.

Sometimes, the answer is your intuition or understanding of the system is wrong and the simulation or measurement is correct. Once you can confirm the result may be correct and your expectation is wrong, this is an opportunity to move up the learning curve and improve your understanding.

It is important to always keep in mind that you can never prove a model or theory or even that your understanding is correct. All you can demonstrate is that a measurement or simulation is consistent with your understanding.

Consistency tests are the most important tests you should always perform. The first and most important consistency test is Rule #9.

Take every opportunity to think of every other consistency test you can. If the indicator light for the 5 V rail is on, it probably means there is 5 V on the rail. Measure the voltage on the rail with a scope to verify the 5 V is present, it is at 5 V, and there is no high-frequency noise on the rail.

If your simulation predicts that the output frequency of the timer is 15.7 kHz, and should be independent of the rail voltage, change the rail voltage by 1 V and verify the frequency is the same, consistent with what you expect.

The more consistency tests you can perform, the higher your confidence that your understanding is close to the reality. Get in the habit of thinking, "If my understanding is correct, what else can I look at to verify the result is consistent with what I would expect?"

There are always far more ways of getting a wrong answer than a correct one, and it is just as easy to screw up a measurement as a simulation. The most important double check we have is consistency tests.

Even after you have performed as many consistency tests as you can think of, it is still no guarantee your answer is correct. It just means you have reduced the risk of an incorrect result.

This process establishes a higher level of confidence in the results.

10.3 Simulating a 555 Circuit

We can build a simple 555 timer circuit in LTSPICE. One of the most common applications of the 555 timer is as an astable vibrator, or a free running oscillator. The TI datasheet for the NE555 timer offers a simple schematic for an astable vibrator, shown in **Figure 10.3**.

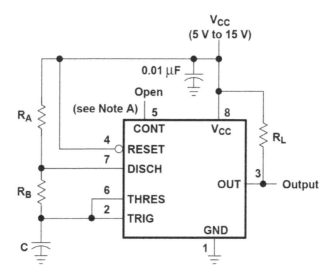

Figure 10.3 Schematic for the 555 as an astable vibrator, from the datasheet.

The model for a 555 timer is in the NE555 component in LTSPICE. The complete circuit and the simulated output voltage on the output pin is shown in **Figure 10.4**.

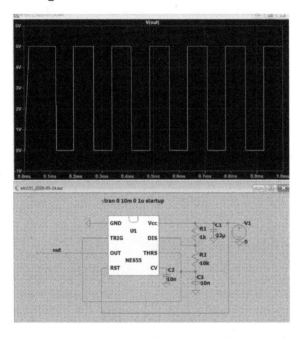

Figure 10.4 Simulated output voltage and SPICE model of a 555 timer circuit as an astable vibrator completed in LTSPICE.

One of the interesting applications of this LTSPICE simulation is to explore the impact of the value of the CV capacitor, connected on pin 5, on the performance of the 555.

The internal schematic of the 555 chip is shown in **Figure 10.5**. The CONT (CONTrol voltage) node is brought out to pin 5. This is part of the resistor divider network that determines the voltage switching levels of the TRIG and THRS levels. Using the voltage divider network with the Vcc as the reference means the threshold switching voltages will scale with the Vcc as will the charging and discharging voltages that determine the timing values. This makes the time to cross a threshold voltage independent of the Vcc fluctuations, and only sensitive to resistor or capacitor values.

10.3 Simulating a 555 Circuit   229

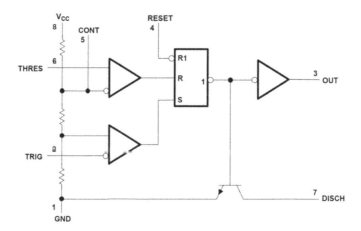

Figure 10.5 Internal schematic and block diagram of the 555 timer inside the chip.

The CONT pin provides the voltage reference to the THRS pin. The datasheet recommends adding a capacitor on this pin to keep the voltage constant. However, it will initially have to be charged up by the internal voltage divider network. Once charged up, it will be very stable.

During the initial charging time, since the voltage on this reference pin will vary, the timing of the output switching will vary as well. In the simulation environment, we can monitor the voltage on the CONT pin charging up while the output is changing.

If the decoupling capacitor is not 10 nF, but 50 nF, the charging time will increase, and this will change the time to initially stabilize the oscillations. **Figure 10.6** shows the time constant to be about 0.2 msec for a C = 50 nF. This makes the effective R charging up the CONT capacitor about 4k ohms.

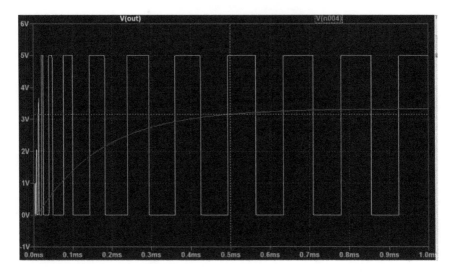

Figure 10.6 Oscillation output (green) while the CONT pin charges up in voltage (red). Once stable, the oscillator output is also stable.

If a 22 uF capacitor was used as the decoupling capacitor, for example, the RC rise time would be 4k x 22 uF = 100 msec. If we use 5 time constants as the time to reach a stable value within 1% of the final value, then it would take 0.1 sec x 5 = 0.5 sec to reach a steady state frequency.

If the application of the circuit required the stable oscillations to start out immediately, the capacitor value selected for the CONT pin would be an important term affecting the circuit response But, if taking 1 sec to reach a final stable oscillation is not an issue, then using a 22 uF decoupling capacitor and minimizing the unique part count on your board would be no problem.

Once we understand the internal circuit of the 555, we can determine the role the CONT pin plays. In principle, there is no need to add a capacitor to the CONT pin. The astable circuit will work just fine if there were no capacitor on the CONT pin. A capacitor will just add a little bit more stability.

This is just one advantage of using a SPICE simulation to explore some of the circuit design and component value trade-offs before building actual hardware.

10.4 Purchase an Evaluation Board

Many specialized ICs are available on evaluation boards. These range from switch mode power supply (SMPS) chips to specialized amplifiers and high-speed drivers.

When selecting a specialized part to test its performance and evaluate the behavior of a reference design, an evaluation board can get you a jump start.

For example, **Figure 10.7** shows a $20 Maxim Integrated SMPS evaluation board with an isolation transformer to enable a floating output.

Figure 10.7 A $20 SMPS evaluation board, courtesy of Maxim.

Many semiconductor vendors provide evaluation boards of their popular IC products. Keep in mind that just because the evaluation board has certain design features is not an indication that these are required. Even when you use an evaluation board from a vendor, always take responsibility for your own design.

10.5 Real Prototypes with Modules

If your design has intelligence in the form of a microcontroller, this means there will be some software development required. The earlier in the product design phase the software can be created and tested, the lower the risk of missing the schedule.

To create and test the software requires a hardware platform with the microcontroller or microprocessor and some of the associated peripherals connected with the bus, as they will be in your final product.

Sometimes, it is possible to build a prototype system out of functional, discrete modules that just plug together in a solderless breadboard before your final custom circuit board is ready.

For example, there are a variety of microcontroller modules that all understand the Arduino IDE, such as shown in **Figure 10.8**.

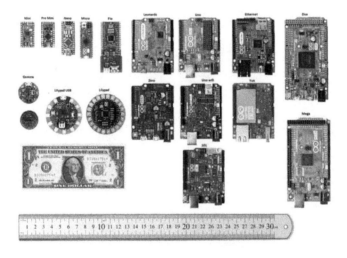

Figure 10.8 A variety of Arduino IDE compatible microcontroller modules to start your modular prototype, courtesy of Electrospeak.com.

There are hundreds of modules that connect to these microcontrollers using standard digital interfaces such as the I2C, SPI, and UART interfaces, or just the digital or analog pins. Examples of some of these modules are shown in **Figure 10.9**.

Figure 10.9 A variety of modules, all with an I2C interface making connection to a microcontroller easy, courtesy of Sparkfun.

If you can build a close approximation of your initial prototype from modular components, you can get a head start on the software development and even user interfaces and identify potential problems early in the design phase.

10.6 Practice Questions

1. What are three reasons to use a SPICE simulation?

2. What is the most important problem to watch out for in a SPICE simulation?

3. What is the most important input needed for an accurate SPICE simulation?

4. Which SPICE tool will you adopt and get familiar with?

5. What is an advantage of a virtual prototype over a real prototype?

6. What is an advantage of a real prototype over a virtual prototype?

7. Pick a microcontroller you are currently using or plan to use. Can you find an evaluation module you can use for software development to design your custom board?

8. What peripheral sensors can you find that can plug into digital pins of a microcontroller module?

9. What peripheral actuators can you find that can plug into digital pins of a microcontroller module?

10. Is there a specific component or functional module you need in a current design that you cannot find as a module?

Chapter 11
Risk Reduction: Prototyping with a Solderless Breadboard

Just because a SPICE simulation gives a result consistent with what you expect does not mean your product will work when you build it as a circuit board.

- ✓ Did you use the same components in your simulation as on your board?
- ✓ Are the models of the components used in the SPICE simulation accurate models that correspond to the actual component you are using on your circuit board?
- ✓ Could you even find the SPICE models of your components?
- ✓ How important are the interconnect parasitics, which might not have been included in the SPICE simulation?

Sometimes building a quick prototype of key elements of your circuit and measuring the performance with a scope is a valuable confidence builder and risk reduction path.

If you wait to test your circuit by building the complete printed circuit board in all of its final complexity, you may introduce a small problem, like tying an enable pin low instead of high or forgetting to connect a ground to your power jack. These will be hard errors, preventing your board from working. They may be either difficult to find, or impractical to repair and require a complete board respin. This means potentially a delay of 2 weeks in your schedule to fix this error, redesign the board, and get it back, only to find another error.

If it is possible to build key elements of your circuit in a solderless breadboard, you can test and evaluate your circuit design in less than one day and save potentially two to four weeks of development time. This is an important risk reduction step.

11.1 Build a Real Prototype

Bob Pease (1940-2011), was a famous analog engineer who used to work at National Semiconductor and wrote many of their op amp application notes, still referenced more than 50 years later. He is shown in **Figure 11.1**.

Figure 11.1 Bob Pease, showing off one of his circuit boards.

He used to say, "My favorite programming language is solder." This was immortalized on a one of his circuit board projects, shown in **Figure 11.2**.

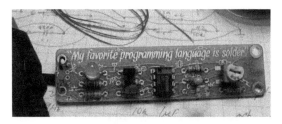

Figure 11.2 Bob Pease used to recommend the approach of building a circuit to experiment.

In his extreme perspective, he did not always trust a circuit simulator like SPICE. After all, the simulated response of the circuit is only as good as the quality of the models used for all the components. It is in the context of using simulations that the phrase garbage in, garbage out originated.

It is easy to get an answer using a circuit simulation tool. But there are many ways artifacts and errors can be introduced. How accurate are the models used for each active component? Are you using the same component in the SPICE simulation as on your board? Did you type in the correct values of the parameters? Did you include the entire circuit? Are your interconnects transparent? If not, did you include accurate models for the interconnect parasitics?

Even with these concerns, by using best simulation practices and by practicing safe simulation, it is possible to get accurate and valuable results from a circuit simulation that correlate well with a measurement. Simulation of a virtual prototype can be an important risk reduction step.

There is also lot of value in building a real, physical circuit prototype to test it out and quickly try different components, different circuit elements, or even different circuit designs.

It is important to keep in mind that it is just as easy to introduce an artifact when building and measuring a circuit as when performing a circuit simulation. Both approaches can give you a misleading result that might steer you in the wrong direction in your final design.

Both simulating a circuit as a virtual prototype and building a physical circuit as a real prototype each have value and they each can be misleading. You have to be careful no matter which path you take in prototyping the circuit.

When used correctly, both approaches can help reduce the surprises and the risk that your final printed circuit board design may not work as expected.

In a physical prototype, you may see unexpected behavior or undocumented features of the component. You can quickly change out components and see the impact on waveforms. The process of building the prototype will help you learn the correct way of using the component. This is part of the qualification process when using a new part.

Many times the datasheet does not contain information about the rise time of the output signals or the source impedance of the driver. This information can unfortunately sometimes only be obtained from measurements on a prototype or evaluation board.

11.2   Solderless Breadboards for POC

The fastest way of building a prototype is using a solderless breadboard substrate. It's just a matter of plugging in components and wiring them up to get the correct connectivity.

In many new and untested designs, you want to be able to establish the circuit design with the components you selected will work/behave as you expect. This is often referred to as the proof-of-concept (POC) design.

If the solderless breadboard is mounted on top of a copper clad circuit board, it can act as a platform onto which other large or mechanical parts can be glued and then wired into the breadboard.

An example of a prototype built in a solderless breadboard, mounted to a copper clad bare circuit board with multiple resistor trimmer pots, connectors, switches, and even BNC connections, is shown in **Figure 11.3**.

Figure 11.3 An example of a solderless breadboard mounted to a bare, copper clad circuit board substrate with multiple peripheral components soldered and wired in.

If the interconnects are transparent, the design will work in a solderless breadboard, providing the schematic design is correct, the components perform as expected, and the connectivity has been implemented correctly.

11.3 Features of a Solderless Breadboard

A solderless breadboard (SBB) is incredibly valuable when prototyping components. It is the base for plugging in leaded electronic components to provide instant connectivity. Many circuits can be prototyped in minutes with an SBB, saving you weeks of potential respin time.

While they come in a variety of form factors, the most useful SBB options are the solderless breadboards that have two vertical columns of holes on either side with many rows of groups of five horizontal pins. The pitch between hole centers is 100 mils (2.54 mm). Inside the holes are small metal spring clips into which we insert leads. These leads are electrically connected together by row or by column.

This pattern is designed to connect dual inline packages (DIPs) with 100 mil center pins and typically 300 mil to 500 mil between columns of pins. An example of an 8-pin DIP in a SBB is shown in **Figure 11.4**.

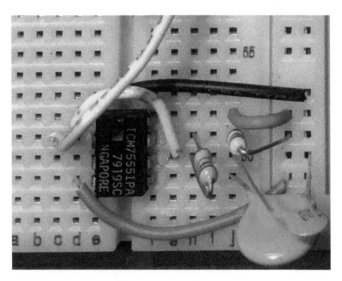

Figure 11.4 An 8-pin DIP in a solderless breadboard.

SBBs usually come in two sizes, full-size and half-size. Examples of these two sizes are shown in **Figure 11.5**.

11.3 Features of a Solderless Breadboard 241

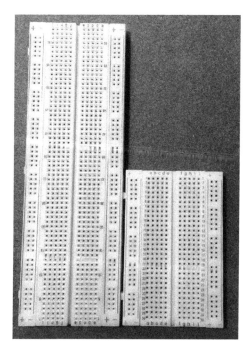

Figure 11.5 Examples of the two sizes of solderless breadboards, full-size (left) and half-size (right).

They all work exactly the same way:

- ✓ All the holes in each vertical column on the outer edges are connected together and adjacent vertical columns are isolated from each other.

- ✓ The five adjacent holes in each group of horizontal rows are connected together, with adjacent rows isolated from each other. The adjacent rows across the middle gap are also isolated from each other.

This connectivity is shown in **Figure 11.6**.

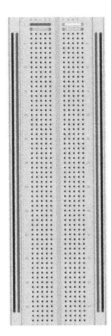

Figure 11.6 A SBB showing the connectivity of the holes. Only a few of the connections between the 5 adjacent holes in groups of rows are shown.

Of course, it is easy to verify the connections using a DMM set for an ohmmeter with the buzzer turned on. In this mode, if the resistance between the leads drops below 200 ohms, the buzzer will buzz. It is always important to get in the habit of verifying specifications. This skill is called reverse engineering a component or a tool.

Some full-size SBBs are built literally from two half-size SBBs. This means the top half of the columns is not actually connected to the bottom half of the columns. Sometimes there is a space in the middle of the SBB, sometimes, not. If the top half is not connected to the bottom half, and you want to use the full size of the substrate, you will have to provide jumpers between the columns. An example of doing this is shown in **Figure 11.7**.

11.3 Features of a Solderless Breadboard

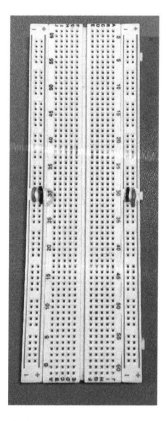

Figure 11.7 Example of a solderless breadboard with vertical columns isolated between the top half and bottom half and the need for jumper wires between them, as shown.

> *Before you use a full-size solderless breadboard, verify the top power rails are connected to the bottom power rails. This is a common source of problems if you are not careful.*

This is why verifying the connectivity with an ohmmeter, especially for the vertical columns from the top to the bottom half of the board, is so important.

There are a number of really great tutorials and descriptions of the solderless breadboard and using it as a connectivity substrate. See for example this one from Sparkfun and this one from SEEED.

The first step in using a solderless breadboard is to design for connectivity. We make sure all the leads on all the components are connected together where they should and not connected where they shouldn't.

11.4 Bandwidth Limitations

In principle, any combination of jumper wires between any components that results in the correct connectivity may work. When rise times are longer than about 1 usec, and noise levels as high as 100 mV are acceptable, the interconnects are virtually transparent, and it probably doesn't matter how the interconnects are wired up.

In practice, as signal rise time decrease, there are two problems the interconnect will introduce: loop self-inductance resulting in rise time degradation and loop mutual-inductance resulting in excessive cross talk.

Every interconnect in every substrate is really two interconnects, a signal path (or power path) and its return. When interconnects are not transparent, it is important to get in the habit of seeing the return path for every signal or power. If we wish to connect a signal from one end of a solderless breadboard to the other, we also have to connect the return. Sometimes we use the vertical column assigned to ground, and sometimes the return is another wire.

An important electrical property of any wire is its inductance. As a good, first-order approximation, the inductance associated with one wire that is part of a signal-return path loop is about 20 nH/inch of wire length.

This simple rule of thumb is based on the loop inductance of a circular loop. Using 24 AWG wire, a loop 1 inch in diameter has a loop inductance of about 64 nH. With a circumference of 3.14 inches, this is about 20 nH/inch of length around the circumference. Of course this is just a rough estimate, but a useful starting place when all that is needed is a rough estimate.

When you see a physical wire in a solderless breadboard, see in your engineer's mind's eye an inductor. An 8-inch-long jumper wire is an inductor of about 160 nH. Including its return path, this is a total loop inductance of 320 nH.

This means that when connecting a signal (and its return) from one end of the solderless breadboard to the other, there is a 320 nH inductor in the series path. In an equivalent circuit model, the loop inductance of the signal and return path are lumped together into one inductor rather than explicitly showing an inductor in the signal and an inductor in the return path.

The impact of this inductor will be to slow down the signal's rising and falling edges due to the low-pass filter of the series inductance and resistance of the circuit. Examples of these long jumper wire interconnects and their equivalent circuit models, including the source resistance and the receiver load resistance, is shown in **Figure 11.8**.

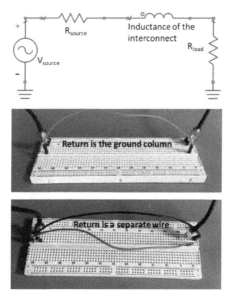

Figure 11.8 Example of a solderless breadboard with jumper wires and the equivalent circuit model.

Between the source and the receiver load is a low-pass filter created by the source impedance and the interconnect inductance. This means that the signal, in traveling across the interconnects, will be low-pass-filtered.

By connecting two ports of a network analyzer between the two ends of the signal-return path interconnects, we can measure the transfer function for the special case of 50 ohm loads on the ends. Figure 11.9 shows the measured transfer function as the frequency of the signal was varied, compared to an ideal model based on the equivalent circuit model with 50 ohm source and load resistances and a 320 nH loop inductor.

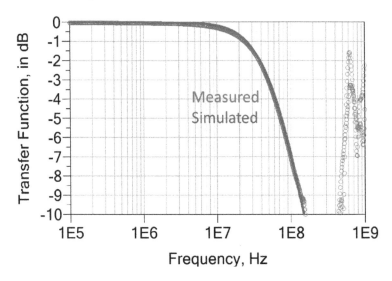

Figure 11.9 Measured transfer function of a long signal and return path either with the ground column or separate return wire compared with the low-pass filter model and 320 nH of loop inductance.

Using the -3 dB value as a rough measure of the highest frequency that can be transmitted with acceptable distortion, the measured bandwidth of the interconnect is about 50 MHz. This is the same as the calculated pole frequency of the low-pass filter of

$$f_{pole} = \frac{1}{2\pi}\frac{R}{L} = \frac{1}{2\pi}\frac{100\,\Omega}{320\,\text{nH}} = 50\,\text{MHz}$$

The distortion will also be related to the source resistance of the driver and the load resistance of the receiver. Larger values of resistances will increase the usable frequency or bandwidth. Shorter-length interconnects will reduce the inductance and increase the usable frequency.

The second problem that can arise is cross talk between adjacent signal-return paths. This is due to mutual inductance between the two signal-return path loops.

When current switches through the aggressor loop, some of its magnetic fields created around the victim loop change. These changing magnetic fields induce a voltage on the victim loop. This is driven by the mutual inductance, M, between the two loops. This is a form of cross talk. The magnitude of the cross talk is

$$V_{victim} = M \frac{dI_{aggressor}}{dt}$$

In the special case of when the current is a sine wave, with frequency, f, and the source resistance of the aggressor loop is Z, the ratio of the noise on the victim to the signal on the aggressor is

$$V_{victim} = M \frac{dI_{aggressor}}{dt} = M 2\pi f \frac{V_{aggressor} \cos(2\pi f t)}{Z} \quad \text{and} \quad \frac{V_{victim}}{V_{aggressor}} = \frac{M 2\pi f}{Z}$$

The inductively coupled cross talk noise will increase with frequency. When the return conductor is shared by the two loops, the mutual inductance is the inductance of the return path, about 20 nH/inch of shared path length. If the shared path length is 8 inches long, the mutual inductance, M, between the two loops is 160 nH.

When the return path is shared by both loops, the expected cross talk for the special case of a source resistance of 50 ohms, with 50 Ohms in the victim loop, is,

$$\frac{V_{victim}}{V_{aggressor}} = \frac{M 2\pi f}{Z} = \frac{160 \text{ nH} 2\pi f}{25} = 4\% \times f\,[\text{MHz}]$$

When the return path, usually labeled as ground, is shared and the return currents overlap, we see the maximum cross talk. This special case of cross talk, dominated by the high value of mutual inductance of the shared return path, is referred to as ground bounce.

At 1 MHz, the expected cross talk is estimated as 4%. It approaches 10% at about 2 MHz. An example of the measured and simulated cross talk for two 8-inch-long pairs of interconnects with a shared return and with separate returns is shown in **Figure 11.10**. The cross talk with a shared return path is very close to the 4% expected value at 1 MHz, and approaches 10% at 2 MHz, as expected.

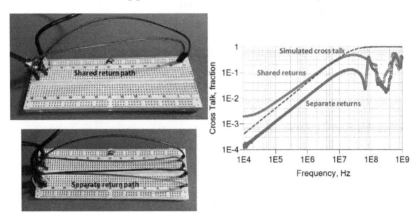

Figure 11.10 The measured cross talk between two loops with and without a shared return path, compared to the simple estimate.

When a separate return path is used for both the aggressor and victim loops, the loop mutual inductance between them is reduced. In the case of separate return paths, the crosstalk approaches 10% at about 10 MHz. When cross talk is the limiting feature of a solderless breadboard, and separate return paths are used, the limiting application frequency is about 10 MHz.

Knowing this limitation, the mutual inductance between two signal-return path loops can be further reduced if the signal and its return path conductors are kept close together with a minimalist loop and far from the victim's signal-return path loop. This will confine the magnetic fields of the aggressor in its vicinity and minimize how many of the field lines couple to the victim.

11.4 Bandwidth Limitations   249

A practical way of implementing this tight coupling between the signal and return paths is by keeping the signal-return path conductors connected together on the ribbon cable. When routing a signal path from one end of the board to another, keeping the return connected to the signal jumper wire will dramatically reduce the cross talk. This configuration and the resulting cross talk between the signal and return loops of an aggressor and victim are shown in **Figure 11.11**.

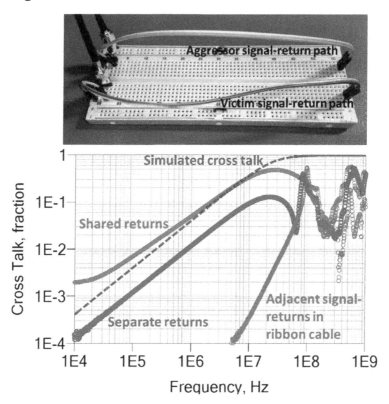

Figure 11.11 The measured cross talk between two signal-return loops still connected to the ribbon cable is shown for various configurations. It is much lower when the return conductors are separate conductors and form a smaller loop area.

This example illustrates the relative cross talk between signal-return path pairs of the same length but differing only in how they are routed. The three examples, in order of largest cross talk to smallest, were:

- ✓ Shared return paths (ground bounce)
- ✓ Separate return paths
- ✓ Signal-return still connected to the ribbon-cable

Using this method of keeping the signal and return paths tightly coupled can increase the application frequency until the cross talk reaches 10% to above 50 MHz.

This analysis suggests that if attention is not paid to how the wiring is done in a solderless breadboard, the application bandwidth may be limited to below 1 MHz before there is excessive noise in the form of cross talk. If good design guidelines are followed, this might be pushed to about 50 MHz.

If low-noise operation above 50 MHz is needed, do not use a solderless breadboard. This is a strong motivation to switch to a printed circuit board interconnect, following best design practices.

The best design practices for routing low-noise signal-return path loops in a solderless breadboard are:

- ✓ Do not share return paths (eliminate ground bounce)
- ✓ Keep the signal-return paths in close proximity
- ✓ Keep the signal paths short

These design guidelines also apply to printed circuit boards, with one important addition: use a wide, continuous plane as the return conductor to reduce the mutual inductance between adjacent signal-return paths. This is a feature not possible in a solderless breadboard but is a very practical feature in a printed circuit board. It is discussed in the next chapter.

> *Watch this video and I will show you a demonstration of the switching noise occurring with long jumper wires when the return paths are worst-case and then optimized.*

11.5 A Simple Breakout Board

To use a solderless breadboard with 100 mil center holes, all the components to be inserted must have pins on 100 mil centers or flexible leads. Some IC components are only available as surface-mount parts. To use one of these parts in a solderless breadboard requires they be soldered to an interface or breakout board that transforms their fine pitch surface-mount leads into 100 mil center leads in two rows.

There are a number of vendors who supply standard format breakout boards that are really geometry transformers. They transform a SMT part into a through-hole part. For example, Bellin Dynamic Systems (http://www.beldynsys.com/products.htm), offers an assortment of transformer boards. An example is shown in **Figure 11.12**.

Figure 11.12 An example of a variety of snap-apart breakout boards for multiple SMT to through-hole parts, courtesy of Bellin Dynamic Systems.

Individual boards can also be purchased from popular suppliers such as Sparkfun, as shown in **Figure 11.13**.

Figure 11.13 An example of a geometry transformer board from an SMT part to a DIP part, available from Sparkfun.

The center-to-center spacing between the rows of holes on either side of the central spacer in a solderless breadboard is 300 mils. This means the minimum spacing between the two columns of pins on opposite sides of the breakout board is 300 mils. It can be increased in multiples of 100 mils, such as 400 mil centers, 500 mils centers, 600 mils centers, etc.

There are only 5 rows of pins on either side of the solderless breadboard. This means that in order to have 1 pin open on either side to connect with other holes, the maximum width board that should be used to transform the fine pitch geometry to the 100 mil center is 300 mils + 600 mils = 0.9 inches. Any larger in width between the two columns of pins and there will be no room for breakout connections. A spacing of 700 mils between header columns enables two open pins on either side of the breakout board for connectivity.

If you plan to use a new part in a circuit, especially if it is a complicated circuit or one for which you are not sure what the optimized component values are, there is value in designing a simple breakout board for the part. You could even include local decoupling capacitors and other simple passives or active components. An example of a breakout board that can be connected up in a solderless breadboard is shown in **Figure 11.14**.

11.5 A Simple Breakout Board

Figure 11.14 An example of two breakout boards to evaluate new components that will plug into a solderless breadboard, designed by my students Kailey Shara and Prem Shah.

This sort of breakout board enables you to incorporate the new part in a solderless breadboard prototype. Using a small eval board will give you a chance to test the footprint and the performance of the part before you design the entire board. This way you can get comfortable with the performance of the part and reverse engineer any important properties not found in the datasheet.

To optimize the performance of the circuit in the solderless breadboard, keep all high-speed interconnects on the circuit board. Route signals for measurement to the external pins. Use best design practices for the breakout circuit board.

When the breakout board with header pins, separated by 700 mils, is inserted into the solderless breadboard, there are two holes on either side available for making connections.

When inserting the breakout board into a solderless breadboard, 24 total pins is about the maximum that can be inserted and removed easily without requiring a screwdriver or pliers.

Of course, the noise generated in a solderless breadboard, even in the best case, will never be as low as in a printed circuit board. But if rise times are longer than 100 nsec or bandwidths less than 10 MHz, the performance in a solderless breadboard may be a good indication of the performance of the circuit if built in a printed circuit board.

11.6 The Mini Solderless Breadboard

There is one other form factor for the SBB that deserves special mention. The previous two form factors of the half-size and full-size are great for building stand-alone circuits. These can be either a simple SBB or an SBB that can mount on top of a small circuit board with header pins to match an Arduino Uno form factor. This can be used to create an instant shield.

An example of an instant shield <u>available from here</u> for only $2 is shown in **Figure 11.15**.

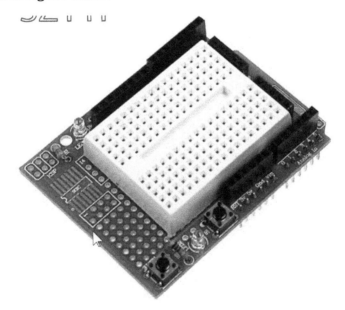

Figure 11.15 Example of the instant shield that can plug directly into an Arduino Uno form factor board.

If you want to build a small circuit that will plug into an Arduino Uno form factor board, this mini, solderless breadboard, mounted onto the circuit board, is the fastest way to build a prototype. It does not have power rail columns along its sides, just the groups of five connected pins in a row. An example of a mini SBB mounted on top of a solderless breadboard header is shown in **Figure 11.16**.

Figure 11.16 An example of a mini SBB mounted onto an Arduino shield header and plugged into an Arduino Uno with a 16-bit, 4-channel ADC and AD584 precision voltage reference chip plugged in.

11.7 Best Wiring Habits

If you use more than two wires in your solderless breadboard, it will quickly get complicated and difficult to keep track of what is connected where. A perfect example of the complexity that can quickly develop in a SBB is shown in **Figure 11.17**.

256 *Practical Guide to Prototype Breadboard and PCB Design*

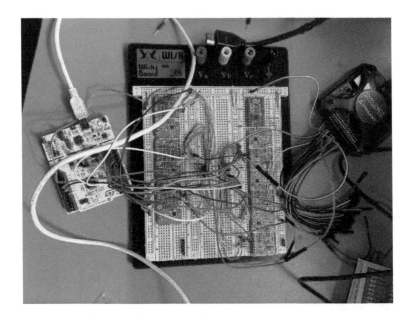

Figure 11.17 An example of a SBB with many modules and jumper wires all over the board. This is very difficult to debug and has some cross talk problems.

To help wire up the board initially and debug or troubleshoot the potential of an incorrect wiring connection, there are a few habits you should always follow.

These simple habits will dramatically reduce the complexity and confusion of your SBB projects and reduce the risk of making a wiring mistake. In addition, they will also help reduce signal quality and cross-talk noise problems.

11.8 Habit #1: Consistent Column Assignments

Habit #1 is to use a consistent set of rules to assign voltage to the vertical columns. Once you pick a rule, follow it for all of your solderless breadboard projects.

The two columns of connected pins on either edge of the SBB should be used to distribute power and ground. Some SBBs label the two columns with a + and - . This designation is completely arbitrary and was done by whoever created the artwork for the SBB, not necessarily because it is a best design practice.

Of course, all connected holes are equivalent. There is nothing special about one column being a + column and one column being a − column. They all behave exactly the same. I recommend you ignore these labels and adopt your own consistent column assignment.

As a general rule, if you are consistent in how you assign the columns, you will make less errors connecting the wrong wires to each of the two columns.

One approach is to use the outer column for ground and the inner column for power. This means that the center-to-center distance between the outer hole in a row and the outer ground column is 400 mils. This is a stretch to span with the signal pin and spring ground lead of a 10x scope probe, but is possible, as shown in **Figure 11.18**.

Figure 11.18 An example of a SBB using the outer column for ground with a 10x probe connecting to a signal output and the ground column for a Teensy microcontroller board.

Using the outer column assigned as ground is arbitrary. It is possible to make a shorter ground connection for probing that will reduce the loop area of the tip using the inner row for ground.

If the inner column on both sides is always used for the ground connection, then a 4-pin header can be used to interface between an SMA connector and coax cable. This will provide a high bandwidth connection between the signals from the solderless breadboard and the 50 ohm coax environment of the cable and the scope. An example of this connection is shown in **Figure 11.19**.

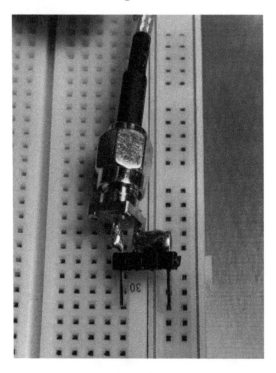

Figure 11.19 An example of a 4-pin header modified to solder to the end of an SMA connector for direct coax cable connection to a solderless breadboard.

This makes for convenient probing at the highest bandwidth. Of course, if the outer column is ground, the same connection can be made but would require a 5-pin header and make the tip with a larger area.

Generally, probing with the shortest ground return connections is easier if the inner column is used for ground and the outer column for power, if it is convenient.

If your board uses just +5 V, the outer column on the right-hand side can be +5 V. The outer column on the left-hand side can be left unconnected, or if needed, also distribute +5 V.

If the circuit also uses a negative rail, the left hand, outer column can be used to distribute the negative rail power. Many bipolar op-amps, for example, have their +Vcc pins on the right side and the -Vcc pins on the left side of the package.

But, in all cases, be consistent and use color coding of the power and ground wires. If you use the outer two columns as power and the inner two columns as ground, at a glance, you can always verify that a red wire going to an outer column is connected to power. A black wire going to an inner column is connected to ground.

11.9 Habit #2: Color Code the Wires

With multiple wires spanning sometimes multiple solderless breadboards, the wiring can get confusing very quickly. An important habit to reduce the confusion is to use particular color-wires for particular applications.

Here is a commonly used color-coding scheme for wiring in general and solderless breadboards in particular:

Red: positive power, typically 5 V

Black: ground

Blue: negative power rail

White: clock lines

Yellow, grey, brown : **digital lines**

Green: analog signals, or sometimes ground

This color code applied to a solderless breadboard is shown in **Figure 11.20**.

Figure 11.20 A good habit is to use the outer two columns for ground connections (black or green) and the inner columns as +power (red), -power (blue) with digital lines as yellow or white or grey.

11.10 Habit #3: Keep Signal Traces Short

Try to use short-length wires from one hole to another. This makes for a neater board, easier to debug, and reduces some signal integrity problems. To reduce the cross talk even more, all long interconnects should be routed as signal-return pairs.

The best way of doing this is using custom length wires from a roll of wire. An example of an assortment of colored solid core wires, available on Amazon for $18, is shown in **Figure 11.21**.

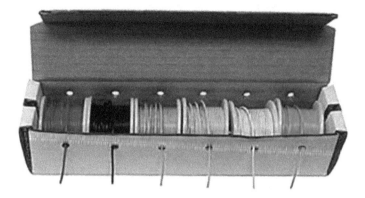

Figure 11.21 Example of six assorted colors of AWG 22 solid core tinned copper wires.

There are two wire gauges that are generally used, AWG 22 and 24. The thicker wire, AWG 22, is ok, but makes a snug fit in the holes of the solderless breadboard. It sometimes requires pliers to insert into the holes in the solderless breadboard. I recommend using AWG 24-gauge wire.

These wires can be cut to a custom length for every specific application. You can use a simple trick to quickly strip the insulation off the ends of the wire to plug them into the holes of the SBB.

Grab the wire about ½ inch from the end with a pair of needle-nose pliers. Hold the wire tight. Use a sharp pair of wire snippers to gently press down on the last ¼ inch of the wire. Don't clamp down so hard as to cut through the wire, but just partway through the insulation.

Gripping the needle-nose pliers tightly, pull the wire snippers away from you to strip the insulation off the end of the wire. The starting place position and ending position are shown in **Figure 11.22**.

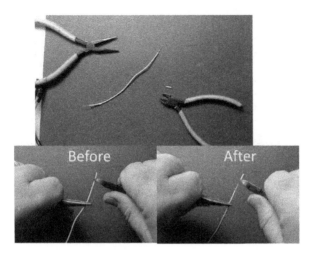

Figure 11.22 Using needle-nose pliers, wire snippers and a simple technique, stripping the insulation off the ends of a wire is very quick and easy.

This takes a little practice to get a feel for the right amount of pressure that nicks the insulation and pulls it off but does not nick and cut the wire.

If you practice this technique, you will be able to trim a wire to the right length and strip the insulation off the ends in a few seconds. This will dramatically speed up how quickly you can wire up a solderless breadboard.

It is generally not a good idea to use large floppy jumper wires to connect one lead to another. This will make for a complicated and hard-to-debug circuit and needlessly increase the loop self- and mutual-inductances between signal-return path pairs, increasing the noise.

Watch this video and I will show you how easy it is to create short wires as needed to connect components in a solderless breadboard.

11.11 Habit #4: Avoid a Shared Return Path

The highest mutual loop inductance between an aggressor and victim signal-return path loop is when the return paths are shared between them. This is the case, for example, when the inner column of holes is used to distribute ground. In this case, there will be ground bounce.

This is certainly the most convenient way of routing signals and their returns. If you use this approach, the inductively coupled cross talk can approach 10% at 1 MHz or above unless switching currents are kept low.

Be aware of the noise that can be generated so that when a low noise design is required, you can implement separate returns and preferably tightly coupled signal-return path loops.

Many solderless breadboard designs are low-frequency applications, and it is an acceptable trade-off to go for convenience. This is the case when signal rise times are longer than 1 usec or low currents are used.

Regardless, when using the common return path of the inner column of conductors, be mindful of the potential impact from ground bounce.

11.12 Habit #5: Route Signal-Return Pairs

If you cannot keep interconnect wires short and must span more than 5 inches, the most important design principle to follow in a solderless breadboard is to use pairs of wires: signal and return paths kept close together.

All interconnect wires should be used in pairs, one wire a signal path and the other its return path, usually ground, unless the signal is a differential signal. These two wires should be routed as physically close together as possible.

This means the two wires connecting between two nodes in your circuit should be either:

- ✓ Still attached in the ribbon cable
- ✓ Twisted together

One wire should be the signal wire and the other its return or ground wire.

If the connecting wires are to a differential pair such as the input to an instrumentation amplifier, or the input to a scope that measured the true differential input signal, such as the Analog Discovery 2 (AD2) scope, keep the two wires still connected on the ribbon cable or twist them together, such as shown in **Figure 11.23.**

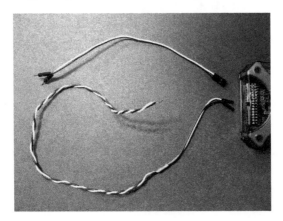

Figure 11.23 The right ways to route the two wires of the differential pair to the Digilent AD2 scope, either still attached to the ribbon wire or twisted together.

11.12 Habit #5: Route Signal-Return Pairs

This principle of routing signal and return conductors in close proximity is the most important physical design principle for all interconnects. It will not fix all noise problems, but it will dramatically reduce inductively coupled noise.

When routing wires board to board, it is especially important to use the signal and return paths in close proximity.

If it is not practical to route the two wires together, then keep them short. In a solderless breadboard, avoid using long floppy jumper wires when you can use short wires that are close to the surface of the solderless breadboard.

Figure 11.24 shows examples of good and bad wiring techniques that will influence the signal integrity problems as rise times decrease.

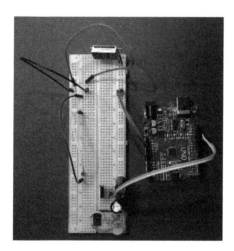

Figure 11.24 Examples of good and bad wiring habits. Good habits are short wires or signals and their returns in close proximity. Bad habits are long, floppy individual wires. They have a return, but the returns are routed farther away, through the solderless breadboard traces.

Be aware of which conductors in the solderless breadboard might be carrying the return current and try to route the signals close to these conductors inside the solderless breadboard.

If you are going to pay attention to one good wiring habit that will reduce the noise generated by the interconnect, it is this one. It adds no cost to a circuit and will get you thinking about some of the hidden, signal integrity problems your interconnect wires can introduce.

11.13 Habit #6: Keep Component Leads Short

When many components such as resistors and capacitors are added to a solderless breadboard, it can sometimes get complicated. There is a tendency to use the long leads that come with the parts and when inserted, make the parts float in space.

Instead, to keep the design neat and orderly, trim the leads of the capacitors so that they are in close proximity to the surface of the board.

For resistors, trim the leads so they are also close to the surface of the solderless breadboard.

When a resistor is spanning just two adjacent holes, instead of trying to keep the resistor body horizontal, rotate it 90 degrees and align the body vertically. This means the two leads will be different heights. They can be quickly trimmed to make the leads the same length for easy insertion into the solderless breadboard sockets.

This makes for a more compact, easier-to-debug circuit design. **Figure 11.25** shows examples of good and not-as-good ways of incorporating leaded parts into a solderless breadboard design.

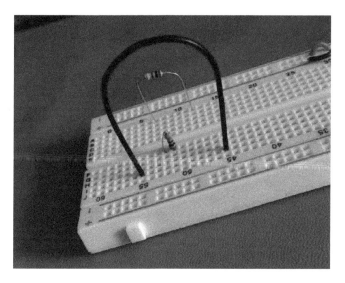

Figure 11.25 Examples of good and bad lead geometries in a solderless breadboard. Where possible, keep leads short to reduce series loop inductance and cross talk. It also makes it easier to follow the connections and debug your circuit.

11.14 Practice Questions

1. What is the difference between a real prototype and a virtual prototype?

2. At roughly what rise time might interconnects be transparent, meaning you may not have to worry about how you interconnect components?

3. In a solderless breadboard (SBB), and generally all connections to instruments, what color should be used for the ground or low-voltage power connections?

4. What color should be used for the higher voltage power connections?

5. When you first get your solderless breadboard, what is the first connectivity feature you should check for?

6. As the rise time of your signals decrease, what is the first problem that might arise in a solderless breadboard because the interconnects are not transparent?

7. When your rise times are shorter than about 1 usec, what is an important design guideline to follow to reduce switching noise?

8. Why is it a good idea to keep component leads short?

9. What is a fundamental problem when using one of the vertical columns to distribute power and ground connections?

10. If you have a sensitive signal line and want to connect it to a detector, what is one thing you could do to reduce the ground bounce noise it might see?

Chapter 12
Switching Noise and Return Path Routing

For DC signals, the resistance of the interconnects in a circuit board or discrete wires in a solderless breadboard will affect signals if the circuit is sensitive to a series resistance of less than about 0.5 ohm.

For signals that are changing, a new source of noise will arise that can sometimes be large enough to cause circuit failures.

The noise arises when signals change, typically when they switch state from a HIGH to a LOW or LOW to HIGH. This type of noise is often referred to as *switching noise*. It will get larger with shorter rise or fall times. It occurs in two types of interconnects, signal paths and power paths.

It is absolutely the case that if you do not take special care in the design of interconnects, switching noise can become so large as to cause failures when rise or fall times are sufficiently short.

An example of the switching noise on a signal line is shown in **Figure 12.1**. This shows the noise on a LOW output of a microcontroller pin, when other I/Os are switching. For each pulse, a larger number of I/Os are switching simultaneously.

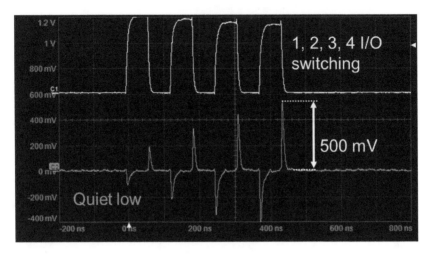

Figure 12.1 An example of the measured switching noise on a LOW output pin when other pins are switching.

In this example, the noise on one nominally LOW pin is 500 mV. For a 5 V receiver, this may not be a problem. But, for a receiver operating at 2.4 V, a voltage above 0.4 V may fool the input into switching. The noise on the output LOW will cause a false switch, which would result in an error and the product would fail.

In busses that are asynchronous, like SPI and I2C, it is possible for noise from other I/O switching to appear during the middle of a bit period when a input on these busses is looking for a 1 or 0 signal level.

Even worse, it would fail only on certain bit patterns on the various I/O and so may be difficult to debug.

By understanding the origin of switching noise we can engineer interconnects in solderless breadboard circuits and in printed circuit boards to reduce the switching noise and increase the useful range of interconnects into higher signal bandwidths.

There is a fundamental limit to how low the switching noise can be engineered using the discrete wires in a solderless breadboard. When we approach this limit, it is necessary to switch to a printed circuit board for the interconnects. The continuous plane and close proximity of signal traces to the return plane in a printed circuit board structure allows an even lower amount of switching noise than a solderless breadboard.

All the principles we develop for the best design practices to reduce switching noise in solderless breadboard interconnects apply directly for circuit board geometries as well.

12.1 The Origin of Switching Noise

Fundamentally, cross talk is due to the coupling of fringe electric and magnetic field lines from an aggressor signal to a victim interconnect. When the interconnects are electrically short, such as when the interconnect length in inches < the rise time in nsec, we can approximate the electromagnetic field coupling in terms of capacitive coupling and inductive coupling.

Switching noise is generated by a combination of the aggressor signals AND the design of the interconnects. The noise coupled onto a victim interconnect is driven by the changing voltage and changing current and is related to:

$$V_{noise} = M \frac{dI_{aggressor}}{dt} \quad \text{and} \quad I_{noise} = C_{coupling} \frac{dV_{aggressor}}{dt}$$

Where

M is the mutual inductance between the aggressor signal-return path and the victim signal-return path.

$C_{coupling}$ is the coupling capacitance between the aggressor signal-return path and the victim signal-return path.

How fast the current or voltage in the aggressor changes is related to the signals. The magnitude of the mutual inductance and coupling capacitance is related to the design of the interconnects.

The signature of switching noise is that the noise on the victim only appears when the aggressor shows a step change in voltage or current. The waveform of the noise on the victim will look like the derivative of the aggressor signal. When the aggressor current is constant, the derivative is 0 and there is no switching noise.

When the aggressor current is a rising edge, the signature of the noise on the victim line is of a pulse with a width comparable to the rise time of the aggressor signal. This is why switching noise only appears on the victim line when the aggressor switches or changes.

For example, **Figure 12.2** shows the measured voltage at the input to channel 2 of a Digilent Analog Discovery 2 scope with channel 2 open, when channel 1 is measuring a square wave signal. The aggressor voltage is 2 V peak to peak. The victim switching noise is 0.02 V peak to peak. This is the signature of capacitive coupling between channel 1 and channel 2 on the circuit board inside the scope.

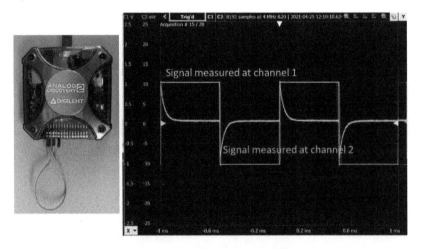

Figure 12.2 Example of switching noise between channel 1 and channel 2 inside a scope, shown to the left. The scale of channel 2 is exactly 1/100th that of channel 1. The switching noise on channel 2 of the scope is 1% the aggressor signal on channel 1.

Capacitive coupling will only show up when channel 2 has a high impedance so that the coupled noise current, flowing through the high impedance of channel 2, such as the 1 Meg ohm input resistance, results in a large, measured voltage. When channel 2 has a 1k ohm resistor between its input and ground, the same capacitively coupled current still passes from channel 1 to channel 2, due to the coupling capacitance, but this current, flowing through the 1k resistor, generates no measurable cross talk signal.

Inductively coupled noise has a similar signature, occurring when there is a changing current. For example, **Figure 12.3** shows a simple case of a digital signal from a microcontroller pin driving a resistor and LED, and then connected to ground. When the current in this circuit switches, it couples inductively to a victim signal-return loop, generating a voltage across it.

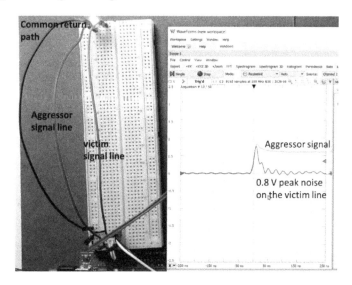

Figure 12.3 An example of an aggressor signal-return loop and victim signal-return loop with a shared return path. The aggressor rise time is about 15 nsec and the cross-talk noise is 0.8 V.

In principle, it is impossible to eliminate all cross talk. There will always be some coupling between aggressor signals and victim interconnects that cause cross talk when signals switch.

In practice, if you follow the best design practices for the interconnects and do everything right, the switching noise can be dramatically reduced.

The lowest cross talk will be when the signal traces use a wide, continuous return plane on the adjacent layer. In this environment, the capacitive and inductive coupling will be of comparable magnitude and both reduced following the same design guidelines:

- ✓ Use a continuous return plane directly underneath the signal traces.
- ✓ Increase the spacing between the aggressor and victim traces as short as practical.
- ✓ Use as short a length over which the aggressor and victim are as close together as possible.

If the return path for the aggressors and victim interconnects is not wide, continuous planes, or if there are gaps in the planes under the signal traces, while both inductive and capacitive coupling will increase, the inductive cross talk will increase much more and is the most important problem to fix.

> *Inductively coupled cross talk is often the dominant root cause of random interrupt failures in microcontroller circuits and should always be the first design problem to fix.*

12.2 Signal-Return Path Loops

All signals travel in a loop on the signal and return path conductors. This is the most important concept in signal integrity.

Unfortunately, the schematic says nothing about the return path. It just shows the connection to ground, as though all ground connections are an infinite sink for current. An example of a typical circuit of a driver, a 60 ohm resistor, and an LED in series is shown in **Figure 12.4**. This circuit says nothing about the return path or where the return current flows.

12.2 Signal-Return Path Loops

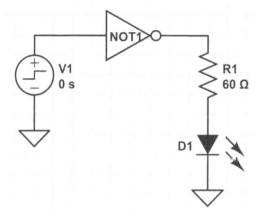

Figure 12.4 A simple circuit with an inverter driving an LED with a current limiting resistor. The schematic says nothing about the signal or return paths.

If the voltage from the pin is 5 V and the resistor is 60 ohms, and the forward voltage drop across the LED is 2 V, the current from the digital pin, through each circuit element is

$$I = \frac{(5V - 2V)}{60\Omega} = 50\text{mA}$$

Where does the 50 mA go when it hits the ground symbol?

When the interconnect is designed, you decide how to connect the return current from the low side of the LED back to the ground connection of the inverter.

In a solderless breadboard, specific jumper wires carry the signal current from the I/O pin to the resistor and LED and then another jumper wire or conductor in the solderless breadboard carries the return current back to the ground connection of the inverter. This is the aggressor signal-return path. An example of this routing is shown in **Figure 12.5**.

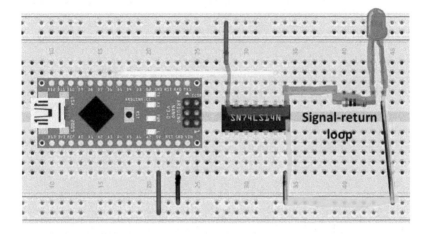

Figure 12.5 An illustration of the signal-return loop created by connecting physical wires between the components.

If there is another connection to an input of the inverter IC with a signal wire and its own return path, this would be a victim signal-return path loop.

If this loop is in the vicinity of the aggressor loop, there will be inductive cross talk on the victim loop. The changing current in the aggressor loop will induce a voltage on the victim loop. Even if the victim signal-return path is shorted together at its far end, there will still be an induced voltage across the short on the victim loop.

The voltage induced and measured between the signal and return loop at the input to the inverter IC will be related to the changing current in the aggressor and the mutual inductance between the two loops.

The mutual inductance between the aggressor loop and the victim loop is related to how many rings of magnetic field lines generated by the aggressor loop also pass through the victim loop per amp of current in the aggressor loop.

12.2 Signal-Return Path Loops

If the current in the aggressor loop doubles, the number of rings of field lines through the victim loop will double, but the ratio of the number of rings of field lines per amp of current will remain the same. Mutual inductance is all about the geometry of the victim and aggressor loops, not about the absolute currents involved.

The fewer the rings of magnetic field lines from the aggressor loop that also circulate around the victim loop per amp of current, the lower the loop mutual inductance and the lower the switching noise.

There are three important geometrical features that will reduce the loop mutual inductance and the switching noise between the aggressor and victim loops:

1. Keep the signal-return loop areas small.
2. Keep the aggressor and victim signal-return loops far apart.
3. Do not share the return path between the aggressor and victim loops.

These principles form the foundation for how to engineer interconnects to reduce switching noise and is the origin of the principle: Don't share return paths, use a continuous plane as the return path. Anyone can see these principles in action using an aggressor loop of wire driven by a square wave from a function generator and a 10x scope probe as the victim loop.

> *Watch this video and I will demonstrate how switching noise is created by mutual inductance between two loops.*

In this example, a function generator was set for a square-wave signal at 10 kHz. The rise time was about 20 nsec. The output was shorted at the far end with leads that were connected together. In this loop, the current was determined by the 50 ohm output resistance of the function generator and the internal Thevenin voltage, set at 5 V peak to peak. This generates a square-wave current of about 100 mA peak to peak, with a 20 nsec rise time.

A 10x probe was used as the victim loop with its signal and return leads shorted together at the tip. The voltage measured across this loop was measured by a scope. It is only during the rising or falling edges of the aggressor current that the changing magnetic fields are generated and the voltage noise, the derivative of the current, is measured.

Figure 12.6 shows how the geometry of the loops affects the switching noise, which occurs only when the current through the aggressor changes.

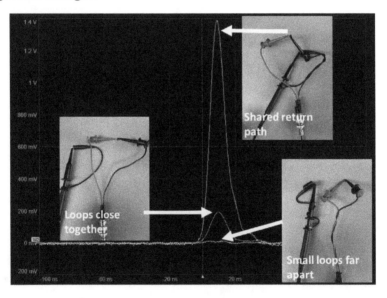

Figure 12.6 Dramatic reduction in switching noise when the loops areas are decreased and moved farther apart compared with a shared return path. All noise measurements are on the same scale.

The advantage of this demonstration is that it does not require any special equipment not normally found in a typical lab and anyone can explore, in a very visual way, the impact from signal-return path geometry on the induced switching noise.

12.3 Where Does Return Current Flow?

If the most important design guideline to reduce ground bounce is to not share return paths, how is it possible for the switching noise to be reduced when a common return *plane* is used for multiple aggressor signals? Isn't this engineering the return currents of multiple signals to share the same conductor?

When the return conductor is a narrow conductor like a discrete wire or another trace on the board, identifying the path the return current takes is very easy. The current is concentrated and distributed roughly uniformly in the narrow return trace. All the return currents sharing this same conductor physically overlap in the narrow return trace.

But, when the return path is a plane, such as a ground plane, where exactly does the return current flow?

In the case of DC current, the actual current path through the plane is a wide path. But at frequencies above about 100 kHz, the high audio frequency end, the path the return current takes in the plane will be directly underneath the signal path and concentrated in the vicinity, underneath the signal current path. As the signal path meanders and wanders over the plane layer in the board, the return current will follow.

It is difficult to calculate the return current distribution in the return plane with pencil and paper as this takes a 3D field solver. But it can be illustrated by a few special cases.

Figure 12.7 shows the cross section of a signal trace over a return plane and the calculated current distribution in both conductors at three different frequencies. The dark blue is no current density and the red means higher current density. Even at 1 MHz, the lowest frequency in this simulation, the current in the return path in the plane is concentrated in close proximity under the signal conductor. This redistribution of the current distribution in the signal and the return path is referred to as the skin effect.

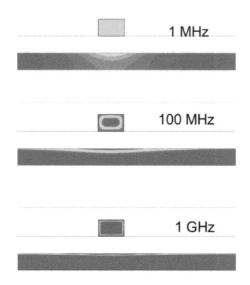

Figure 12.7 The current distribution in the signal and return path of a microstrip transmission line at three different frequencies. Blue means no current, red means highest current density. The return current is located under the signal path at frequencies above 1 MHz.

The skin effect is ultimately driven by the signal and return current taking the path of lowest impedance. At any frequency, the impedance of the signal-return path loop is related to the resistance of the path, R, and the loop inductance of the path, L, by

$$Z = R + j2\pi fL$$

At DC, f = 0 and the impedance of the signal and return path is dominated by DC series resistance, R. The current in the wide plane of the return path will spread out in the entire plane to take the path of lowest resistance. However, as frequency increases, the inductive reactance component of the impedance will increase due to the f term. At some frequency, the inductance will dominate the impedance and the current will redistribute to reduce the loop inductance.

At higher frequency, the inductive reactance dominates the signal-return path impedance, and the current will redistribute to reduce the loop inductance. This means within each conductor, the current will spread out as far as possible from itself. This will reduce the self-inductance of each conductor. The current is usually constrained by the signal trace geometry.

To reduce the loop inductance, the return current will redistribute to be as close as possible to the signal current. This increases the mutual inductance between the signal and return currents, decreasing the loop inductance.

When the inductive reactance term dominates, the return current will always redistribute to keep the loop inductance as low as possible. As the signal conductor routing changes, the return current in the return plane will follow underneath. Keeping the signal and return paths in close proximity will keep the signal-return path loop inductance at a minimum. **Figure 12.8** shows the simulated current in the return plane changing as the signal path changes for three different frequencies.

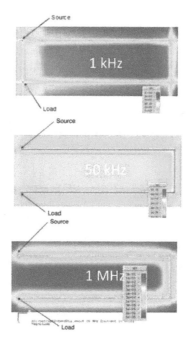

Figure 12.8 Calculated current density in the signal and return path when the signal path changes direction. At 1 kHz, the return current takes the path of lowest resistance. At 50 kHz, most of the return current is under the signal trace. At 1 MHz, all of the return current is under the signal trace. Courtesy of Bruce Archambeault.

In this example, a signal is routed in a U-shaped conductor on top of a plane. The current is launched between the signal and ground at one end of the U and the signal and return are shorted together to the plane at the other end of the U.

At 1 kHz, the return current takes a shortcut between the ends of the U to take the path of lowest resistance. At 50 kHz, the loop inductance dominates the current path and the close proximity of the signal and the return directly underneath is a lower loop inductance than the short cut. And, at 1 MHz, the current is fully under the signal conductor.

When you look at the signal traces on a board that are routed on top of a ground plane, you have to see with your engineer's mind's eye the return paths these return currents follow.

Even though the plane is very wide, the return current is only flowing in a region very close to where the signal current flows.

When there is an adjacent signal path separated by just a few line widths, the return currents from these two adjacent signal traces do not overlap and are separate and succinct. **Figure 12.9** shows the current distribution for two adjacent traces. These two conductors share the same physical return conductor, but above about 100 kHz, their return currents do not overlap and they take separate paths. There is still inductive cross talk, but it is much less than when the return paths overlap.

Figure 12.9 Current distribution in two adjacent signal paths and their separated return current in the plane below.

When the signal trace is routed along some path on the signal layer, the return current will follow underneath in the return plane. Above about 100 kHz, at frequencies where the switching current noise would play a role, the return current in the ground plane will always follow the path of the signal trace and remain directly underneath the signal path.

For another demonstration of the frequency dependence of the return current path, see this article: https://www.signalintegrityjournal.com/articles/1771-a-simple-demonstration-of-where-return-current-flows

12.4 A Plane as a Return Path

Even though a continuous return plane is a shared conductor, because of the properties of return currents to hug the signal trace, the return currents do not overlap in the plane. This is easily demonstrated with a simple test board.

A 2-layer board was designed and built with two sets of six aggressors. One set had an external return trace and the other had a return plane. The six aggressors were all driven simultaneously by digital I/O signals from a microcontroller. The noise on the victim trace was measured. This board, driven by the microcontroller, is shown in **Figure 12.10**.

284 *Practical Guide to Prototype Breadboard and PCB Design*

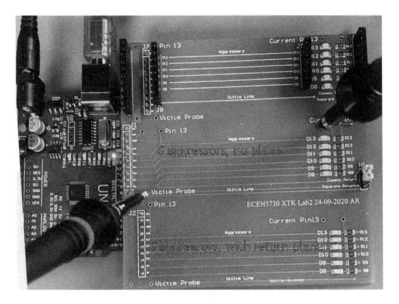

Figure 12.10 The test board with two sets of six aggressors driven by the micro controller.

In the first set of aggressor traces, one option was for a single return trace for all six traces that is shared by the victim. The second option was for a separate return trace for the six aggressors and the one victim line. This section of the board is shown in **Figure 12.11**.

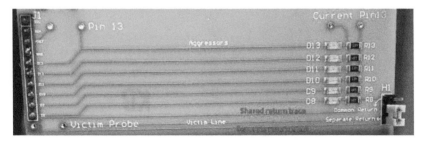

Figure 12.11 The section of the board with six aggressors and one victim line. The return can be selected as a shared return or a separate return for the victim trace.

The six aggressor traces were all driven simultaneously with digital signals from a microcontroller with rise times of about 3 nsec. Each 4

V signal drives a 70 ohm resistor and LED. This is a current change of about (4 V – 2 V) /(70 ohm) = 29 mA per I/O.

All six aggressor signals share the same return trace. When the return trace is also shared by the victim line, the noise on the victim line in this example is 3.5 V peak to peak. This is huge.

When the return of the victim trace was routed as a separate trace adjacent to, but not shared with the aggressor traces, the noise on the victim trace dropped to less than 0.35 V. These measured waveforms are shown in **Figure 12.12**.

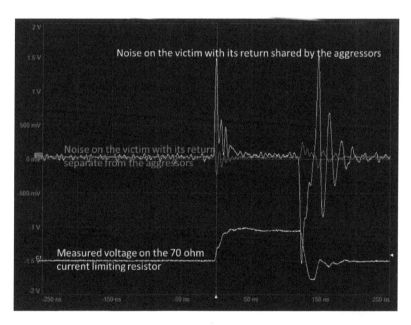

Figure 12.12 The measured voltage noise on the victim line with a shared return and with a separate return.

By not sharing a return trace, but using a separate return, the cross talk between the six aggressors and one victim line was reduced by 10x.

The exact same six aggressor traces were constructed in another board with a continuous return path. This section of the board is shown in **Figure 12.13**.

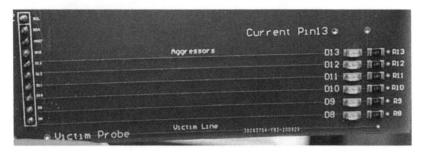

Figure 12.13 Section of the board with six aggressor traces and one victim trace, all with a shared return plane. The victim trace is connected to the return plane.

Under exactly the same conditions of six aggressors switching simultaneously, but with the return currents propagating in the plane beneath the signal traces and the same adjacent return trace, the cross talk noise was measured as less than 0.1 V. This is shown in **Figure 12.14**.

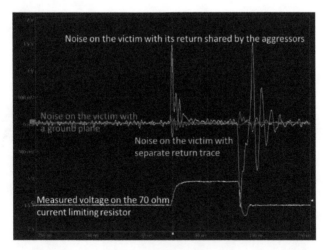

Figure 12.14 Measured noise on the victim for the three cases of victim trace with a shared return, with a separate return, and with the aggressors and victim sharing the same return plane.

When the return path for the aggressors is a continuous plane shared by the victim trace, there is still cross talk, but is it less than 0.1 V when the aggressor signals are 4 V. The use of a return plane brings the cross talk on the victim trace with a shared trace as the return from 3.5 V peak to peak to 0.1 V peak to peak. This is why you should always use a continuous plane as the return path.

> *Watch this video and I will demonstrate the difference in switching noise between a separate, common return trace and a return plane.*

12.5 Ground

The ground net in a circuit is used for two very different purposes.

First, it is the reference node from which all voltages in the circuit are defined or measured. In principle, we assume every point on the ground net is at the same voltage or that the ground net is an *equipotential*. This means it does not matter which specific location on the ground net is used as the reference point. They are all equivalent.

In practice, the ground net is not always an equipotential. There can be voltage differences between one point on the ground net and another. Voltage differences between different locations on the ground net can be created when currents flow through the resistance and inductance of the ground net. This is literally ground bounce noise.

To distinguish between the different voltages on the ground net used as a reference, we refer to *local* ground as the node on the ground net in proximity to the signal node measured relative to that location.

Even though every point on the same net has a DC path to every other point, there may be a voltage difference between different local ground nodes. This may be due to voltage drops from currents flowing through the resistance of the ground conductors or from changing currents flowing through the inductance of the ground conductor. These are examples of ground bounce noise.

Ground bounce in the ground net will introduce noise between different local nodes when the ground net is used as a reference. This is especially important for low noise applications. In these cases, routing the sensitive signal traces far away from the high-current, fast-changing signals or power paths which use the ground net as their return path will reduce the ground bounce noise they see in the ground plane.

Another approach when measuring low-level voltage signals sensitive to ground bounce from other switching currents is to route a reference voltage in a separate conductor along with the sensitive signal trace. This second reference conductor will not carry any current so will not have ground bounce noise. It is a local ground that is a separate conductor than the return path ground.

This is the principle of a differential pair used for analog signals. One conductor connects to the sensitive signal, and the other conductor to the reference voltage to which the sensitive signal is measured where it is created. It is the voltage difference between the two conductors that is the sensitive voltage signal. The use of differential pairs to carry noise-sensitive signals is discussed in a later chapter.

The second purpose of the ground net is to carry the return currents for signals and power. This is a completely different use of the ground net than as a voltage reference.

To distinguish the two applications for the ground net, sometimes the term *digital ground* is used to refer to the return path for the switching digital signals.

The term *analog ground* is used as the return and the reference for the analog signals. Generally, the analog signals do not involve much current and would not create much ground bounce noise in the analog ground paths. Analog signals are also more sensitive to ground bounce noise.

The same ground plane can be used to carry both the digital return currents and the analog return currents. There is rarely a need to use separate analog and digital ground conductors. The way to reduce the digital ground return currents' ground bounce from affecting the analog signals' reference voltages is to route the analog signals away from the digital signals, using the same return plane. The return currents for the digital signals that will create ground bounce, will be in proximity to the digital signals and not spread out over the entire ground plane of the board.

There is rarely a strong compelling reason to split a ground plane to separate digital returns and analog reference grounds.

There is also the ground associated with the power return, but it is not distinguished from digital or analog ground. In addition to analog and digital ground, there are two other types of ground.

Earth ground is a conductor that is tied with a low resistance path to a copper pipe stuck into the literal ground and buried at least four feet deep. This is an important *safety* feature. Most residential and commercial buildings have building codes that specify how large a pipe should be used and how deep it should be set in the ground.

In an AC power plug socket in the wall of your home, the round hole under the two prong holes in U.S. outlets is connected by a thick copper wire to the ground post sticking in the ground outside the building.

Chassis ground is the connection to the metal enclosure of an electronic product. For safety, the chassis ground should be connected to earth ground. This is an Underwriters Laboratory (UL) requirement for many products. This reduces the risk to a consumer in touching a metal surface that might accidently be wired to a higher voltage.

12.6 Avoid Gaps in the Return Plane

The first step in reducing the loop mutual inductance between two signal-return path loops is to identity the *return paths*. Unless you have reason to believe otherwise, assume the return path of a signal line is the adjacent ground path beneath the signal trace.

If you need to route a signal line down to the bottom layer in the ground plane layer to cross under a return plane, there will be a gap in the bottom layer ground plane around the cross under trace. This gap is a discontinuity in any return current trying to flow across the gap. An example of the gap in the return path with a cross-unders is shown in **Figure 12.15**.

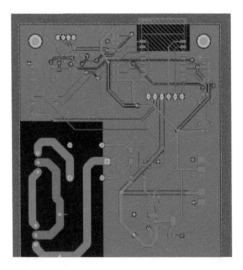

Figure 12.15 A 2-layer board with the bottom ground layer in blue. Note that all cross-unders into the bottom layer create gaps in the return plane.

When the trace on the top layer crosses over the gap in the return path, its return current is forced to flow around the gap, increasing its loop self-inductance. This is illustrated in **Figure 12.16**. This by itself may not be a problem. This signal-return path will just see a larger loop self-inductance.

12.6 Avoid Gaps in the Return Plane

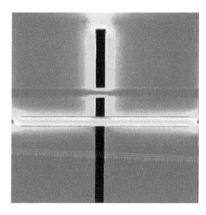

Figure 12.16 Current density in the return plane when an aggressor signal crosses a gap (black) in the bottom plane. The return current has to snake all the way around the gap, increasing the loop self-inductance and loop mutual-inductance to other victim signal-return paths.

However, if there is another adjacent signal trace ALSO passing over the same gap, its return current will also have to snake around the gap, overlapping and having a large mutual inductance with the aggressor trace.

When the return currents of multiple signal paths are forced to overlap due to a gap in the return plane, there will be ground bounce cross talk.

The more signal traces crossing over the same gap in the return path, the larger the loop mutual inductance between the signal-return path loops and the more switching noise on the victim loop.

If you need to add a cross-under in order to route one signal line across the path of another, you will introduce a gap in the return plane.

The way to minimize the impact of the gap on the return path is to keep the length of cross-unders short. Where possible, keep the gap length less than 0.5 inches long. If a long trace has to cross many signal traces, it is better to use multiple short cross-unders than one long cross-under creating a long gap in the ground plane.

If a long cross-under path length is necessary, creating a long gap in the return plane, add a *return strap* to provide a connection for return currents to jump over the gap. This is a short trace that connects the ground plane on either side of the gap using a *cross-over* to the top layer. This return strap will reduce the loop mutual inductance between the two signal lines that pass over the gap. An example of two return straps over a gap in the return plane to provide a continuous return current for the signal traces crossing over the gap is shown in **Figure 12.17**.

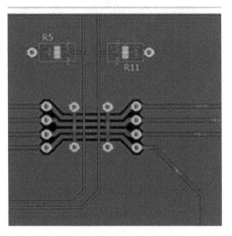

Figure 12.17 An example of a cross-over providing a continuous return current path for signals crossing a gap.

Even if your design is not sensitive to switching noise, it is a good habit to always design for reduced loop mutual inductance.

12.7 Summary of the Best design practices

To reduce the loop mutual inductance between two or more signal-return path loops:

1. Do not share individual traces or wires as return paths.

2. In a 2-layer board, use the bottom layer as a continuous ground-return path.

3. Route the signal paths as far apart or spaced out, as practical.

4. If you need to route a signal line on the ground plane layer, keep its length as short as practical.

5. If there is a gap in the return path longer than 0.5 inches, add a return strap over the gap to provide a continuous current path adjacent to the signal traces crossing over the gap.

6. As a final habit, follow each signal trace. Verify the return path is continuous underneath. Span any discontinuity with a return strap to keep the return path in close proximity to the signal path.

7. Consider using some of the space under the IC footprint as a ground connection with ground vias to the bottom layer, and wide connections to the ground leads of the package, using thermal reliefs to the pads if necessary. These are called ground puddles.

12.8 Practice Questions

1. What are the top three best design practices to reduce cross talk?

2. Where is the return current to a trace when it is routed over a plane?

3. If you route a signal layer to the bottom layer as a cross-under, why is there a gap in the return plane?

4. If the signal trace makes a turn on the top layer, what does the return current do in the ground plane on the bottom layer?

5. What does a return strap do?

6. Suppose there are three signal lines that cross over a gap in the return path. Where is the best place to add the return strap?

7. What are three good guidelines for engineering the return strap across a gap in the ground plane?

8. Why does just adding a copper fill or pour to a top layer do nothing for return path control?

9. Suppose the mutual inductance between two loops is 20 nH. The signal is a 1 V peak to peak square wave signal into a 50 ohm resistor and the rise time is 10 nsec. What is the estimated peak cross talk voltage? What is the current waveform of the aggressor signal, the current waveform, and the cross-talk voltage noise expected on the victim line?

10. Which is a better route that would result in less cross talk: two signal lines with the same return trace or routing the return paths in the same common ground plane on the bottom layer of the board?

11. Suppose you measure 10 mV of cross talk for a 5 nsec rise time and 10 mA transient current in the aggressor. What would you estimate to be the loop mutual inductance between the two loops?

Chapter 13
Power Delivery

Switching noise arises when currents change or switch. The noise is the induced voltage created by a changing current through an inductor.

In Chapter 12, we showed examples of how the changing current in one signal-return path loop can create a voltage noise in an adjacent signal-return path loop. This switching noise is a type of cross talk. It is largest when the return paths of the two loops are shared and smallest when the signal-return path loops are tightly coupled and far from each other.

Another type of switching noise occurs in the power path also referred to as the power rail.

Power rail related switching noise is due to loop inductance in the power distribution path from the voltage regulator module (VRM) to each IC. We call this path the power distribution network (PDN).

This type of switching noise is especially important in solderless breadboard interconnects. It occurs in printed circuit boards, but is easier to reduce.

Switching noise in the PDN is due to the loop inductance in the power-ground path. There will be self-aggression noise when one device draws current from the power rail and mutual-aggression noise if other devices are connected to the power rail and see this noise.

13.1   Origin of Power Rail Switching Noise

A changing current through an ideal inductor creates a voltage drop across it, as shown in **Figure 13.1**.

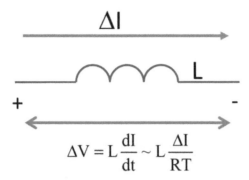

Figure 13.1 The voltage across an inductor is related to how fast the current through it changes.

The interconnects in the power path from the voltage source, the voltage regulator module (VRM), to the power and ground pads of the IC, have some loop inductance.

As the active devices, like the microcontroller, the communications ICs, the logic devices, and even the analog amplifiers, perform their activities, they draw current. As the microcode changes or the operations change, the current draw through the power rail changes.

It is this changing current through the inductance of the power rail that causes power rail switching noise. A voltage drop across the inductance from the VRM to the power pins of the IC means the voltage on the devices' power pins will momentarily drop below the VRM voltage during the dI/dt.

Given the capacitances and inductances in the PDN path, this transient current might also excite some LC resonances causing ringing on the power rail pins of the IC. **Figure 13.2** illustrates this equivalent electrical circuit model of the PDN.

13.1 Origin of Power Rail Switching Noise 297

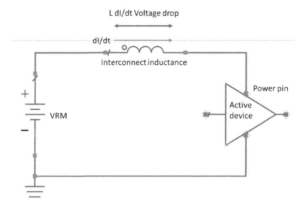

Figure 13.2 Simplified model of the power distribution network (PDN) with the interconnect inductance that causes switching noise.

An example of power rail switching noise measured on the Vcc rail on the die of an IC is shown in **Figure 13.3**. When this voltage drops too low, it will cause errors in the functioning of the device.

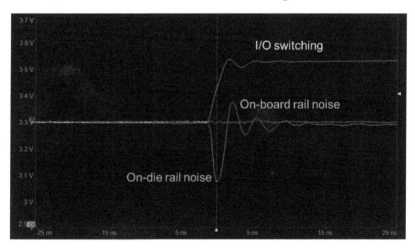

Figure 13.3 Measured voltage noise on the power rail of a digital IC as its I/Os switch, driving current through the power rail. Scale is 100 mV/div. Switching noise is 200 mV. The ringing is from an LC resonance related to the capacitance on the chip and the PDN inductance.

Rarely do we have the opportunity to reduce the changing current from the devices as this is what performs the circuit function. This means the most important way of reducing power rail switching noise is to reduce the loop inductance between the power pins of the IC and the VRM.

13.2 Calculating Loop Inductance

The amount of loop inductance depends on the length of the leads, their separation, and their geometry. The routing geometry of the power and the ground return influences the amount of loop inductance. There are very few structures for which there are exact equations to calculate the loop inductance of a path. Instead, we rely on simple approximations for a few specific, well-defined geometries. Three in particular are useful:

- ✓ A circular loop of wire
  (https://www.allaboutcircuits.com/tools/wire-loop-inductance-calculator/)
- ✓ Two round, parallel circular wires of some length
  (https://www.allaboutcircuits.com/tools/parallel-wire-inductance-calculator)
- ✓ Two coplanar conductor strips
  (https://technick.net/tools/inductance-calculator/coplanar-traces/)

For example, in a power-ground path that is in the shape of a circle, the loop inductance is approximately:

$$L = 32\text{nH}/\text{in} \times \frac{D}{2}\left(\ln\left(\frac{8D}{d}\right) - 2\right)$$

where

D = the loop diameter in inches

d = the wire diameter in inches

These dimensions are illustrated in **Figure 13.4**.

13.2 Calculating Loop Inductance

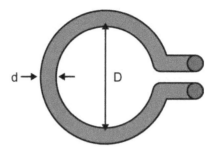

Equations

$$L_{loop} \approx \mu_0 \mu_r \left(\frac{D}{2}\right) \cdot \left[\ln\left(\frac{8 \cdot D}{d}\right) - 2\right]$$

Figure 13.4 The features to calculate the loop inductance of a circular loop. Source is AllAboutCircuits.com.

In the special case of a loop 1 inch in diameter composed of 24 AWG wire, 20 mils in diameter, the loop inductance is,

$$L = 32\text{nH}/\text{in} \times \frac{D}{2}\left(\ln\left(\frac{8D}{d}\right) - 2\right) = 32\text{nH}/\text{in} \times \frac{1}{2}\left(\ln\left(\frac{8}{0.02}\right) - 2\right) = 64\text{nH}$$

The circumference of a 1 inch diameter loop is about 3.1 inches. This makes the loop inductance per inch of the circumference about 64 nH/3.1 inch = 20 nH/inch. As a rough rule of thumb, the loop inductance of the power-ground path is about 20 nH/inch of circumference along the power and the ground paths.

If the power currents, which return on the ground paths, are constant and just DC, then there will be no dI/dt through the inductance of the power path, and there is no switching noise.

However, when large transient currents are drawn by the IC, like when I/Os switch, or LEDs turn on, or a microcontroller comes out of an idle state, the transient current through the power rail's loop inductance can result in a large voltage drop.

To reduce this voltage noise on the power rail, the most important design guideline is to reduce the loop inductance between the pads of the IC and the VRM.

13.3 Measuring PDN Switching Noise

The amount of power rail switching noise depends on the loop inductance from the VRM to the IC power pads and the dI/dt of the switching current. If the dI/dt is small enough, the switching noise may be small enough to not be a problem in your current design.

But guaranteed, one of your next designs will use circuit elements with a shorter rise time or larger switching currents and how you implement the PDN wires will affect the noise generated. It is absolutely guaranteed that at short enough rise times, the interconnect wires, either in a solderless breadboard or in a printed circuit board, will create measurable noise that may get so large as to cause your circuit to fail.

An example of the power line switching noise generated on the 5 V power rail of the exact same circuit, just using a driver drawing current from the power rail with a 0.5 usec rise time and a faster driver with a 0.05 usec rise time, is shown in **Figure 13.5**.

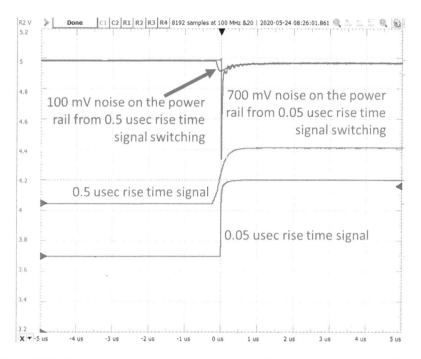

Figure 13.5 The measured noise voltage on a power rail from the same circuit, with the same switching current, but one case using a long rise time and in the other case a shorter rise time. Upper traces are the noise on the power rail. Lower traces as the signals drawing the current.

In this circuit, a long rise time results in a power rail dip of about 0.1 V out of a 5 V rail, or 2%, probably not important. When the rise time is made shorter, the exact same interconnects show a noise voltage on the power rail of a 0.7 V drop out of a 5 V rail. This is a 14% drop, a noticeable and maybe significant problem.

Some of this switching noise can be reduced with good interconnect design. An example of the impact of good and bad wiring on the noise generated on the power line of the 5 V rail is shown in **Figure 13.6**. It is the exact some circuit, with the power rail routed with interconnects having a large loop inductance and a smaller loop inductance.

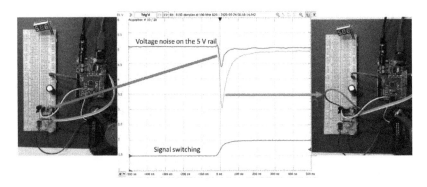

Figure 13.6 Using a short wire routing power to the switching transistor results in a reduction of voltage noise from a 2 V drop to only a 0.5 V drop.

13.4 The Role of Decoupling Capacitors

In practice, it is sometimes difficult to reduce the inductance in the wiring path from the VRM to the Vcc pads of the IC. The distance is often fixed due to the necessary placement of the parts.

The alternative is to add a local source of charge that can flow through the power rail to the active device during the period when there is a transient current, the rise or fall time of the changing current. This is a temporary, local VRM to provide the current at nearly constant voltage only during the critical switching time.

This source of local charge is a capacitor.

A capacitor stores charge at the price of voltage. As a charge reservoir, it can provide the local current needed by the switching device for the short time of the switching current. For a short period of time, it can act as a local VRM. As current flows from the capacitor, its voltage drops. Once charged to a voltage, as current flows out of the capacitor, the voltage drops as

$$\Delta V = C \Delta Q = C \times I \times \Delta t$$

13.4 The Role of Decoupling Capacitors

The capacitor is selected so that during the time in which we expect to see a transient current, the rise or fall time, the voltage drop from the current draw is an acceptably small drop.

We can estimate how much decoupling capacitance we need based on the current draw, the current's transient time, and the acceptable voltage change:

$$C = \frac{\Delta Q}{\Delta V} = \frac{\Delta t \times I}{\Delta V} \tag{1.1}$$

Where:

C = the amount of decoupling capacitance needed

ΔQ – the charge depletion of the capacitor to supply the current requirements of the chip during the transient current time.

ΔV = the acceptable voltage drop (noise) on the capacitor, which is the voltage droop on the capacitor as the current bleeds off charge

Δt = the time during which the transient current occurs

I = the current draw of the die

For the specific case of:

I = 100 mA

ΔV = 3.3 V x 5% = 160 mV

Δt = 10 usec, the longest expected rise or fall time of the transient current

The amount of decoupling capacitance needed is about:

$$C = \frac{\Delta t \times I}{\Delta V} = \frac{10 \text{ usec} \times 100 \text{ mA}}{160 \text{ mV}} = 7 \text{ uF} \tag{1.2}$$

Of course, the precise value of capacitance needed to provide the local charge depends on the amount of time during which the transient current is flowing. We want all this transient current to come from the local capacitor. For a larger transient current, we need a larger capacitance. For a shorter transient time, we need less capacitance.

Unless you put in the numbers for your specific case, a good starting place is to use a 10 uF – 22 uF decoupling capacitor. The precise value of capacitance is not important as long as it is larger than the minimum required to supply the local charge during the switching event. Using a larger value capacitor will have no impact on the switching noise.

As an illustration, a simple switching circuit was built and the switching noise measured for two different values of capacitors, a 1 uF and 1000 uF. **Figure 13.7** shows the measured voltage drop from the inductance between the capacitor and the pads of the IC drawing the switching current for these two cases. Since the inductance did not change, the switching noise is the same, even though the capacitance was increased by a factor of 1000x.

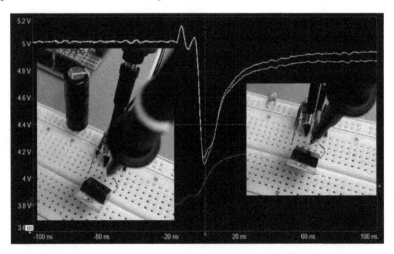

Figure 13.7 Measured switching noise on the power rail of a switching IC with a 1000 uF and a 1 uF decoupling capacitor in close proximity. There is not difference. The switching noise is not about the capacitance as long as it is above a minimum value, it is about the loop inductance between the IC and decoupling capacitor.

The most important quality of the capacitor is the loop inductance between the capacitor and the device to which it provides the charge. The closer the capacitor is to the device, the lower the loop inductance through which the switching currents flow. More capacitors in parallel may reduce the loop inductance associated with the leads of the decoupling capacitor.

The close proximity of the capacitor to the switching device decouples the inductance of the rest of the PDN during the switching time. This is why we call these capacitors that supply local charge decoupling capacitors.

The decoupling capacitor provides current to the local devices, allowing the transient currents to bypass the high inductance of the rest of the PDN. This is why a decoupling capacitor is also sometimes referred to as a *bypass* capacitor.

If there is little dI/dt in the power rail of the IC, it may not need much decoupling capacitance. But, if it is switching 10 A in 1 usec to power an LED strobe, it may require as much as 100 uF of decoupling capacitance.

> *Unless you have a specific requirement for your device, a good starting place is to use at least 10 uF of decoupling capacitance. It is always good practice to do your own estimate.*

This amount of decoupling capacitance will help reduce the voltage noise on the board level power rail from the IC switching currents, before the VRM can bring the voltage droop back up.

13.5 Where Do Decoupling Capacitors Go?

Once the minimum amount of capacitance is selected, the second important property of the decoupling capacitor is its location. During the switching time when the transient current flows, it acts as the local VRM.

To reduce the switching noise on the power pins of the IC, the loop inductance from the capacitor to the IC pins must be reduced as much as possible. This means the decoupling capacitor should be placed as close to the power pins of the IC as possible.

With sufficient capacitance in the decoupling capacitors, there will be little current flowing through the PDN interconnects from the decoupling capacitors to the VRM during the switching time. This means the loop inductance of the PDN from the VRM to the capacitors is not a critical design parameter. It is really ok for this interconnect path to take a longer, higher inductance path. The decoupling capacitor decouples the inductance of the rest of the PDN from the IC.

Once enough capacitance is allocated to provide the switching current, more capacitance has NO impact on reducing the transient switching noise of the power rail. The most important feature of the decoupling capacitors is their proximity to the IC and their low loop inductance.

The loop inductance from the decoupling capacitor to the IC comes from three sources:

- ✓ The loop inductance inside the IC package
- ✓ The loop inductance from the IC pads to the capacitor
- ✓ The equivalent series inductance (ESL) associated with the leads and internal structure of the capacitor

This equivalent circuit is shown in **Figure 13.8**.

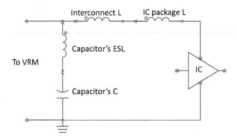

Figure 13.8 Simple equivalent circuit model of the total loop inductance from the pads of the IC to the decoupling capacitor.

13.5 Where Do Decoupling Capacitors Go?

The residual loop inductance from the decoupling capacitor to the IC pads is what generates the residual switching noise. This will appear as a voltage noise on the pads of the IC. This noise will be related to:

$$V_{noise} = L \frac{dI}{dt}$$

To reduce the loop inductance, we need to use capacitors with short leads and in close proximity to the IC. The inductance is the loop inductance of the power AND ground path between the capacitor and the Vcc and gnd leads of the IC.

In a solderless breadboard, it is almost impossible to reduce the path length from the pads of the IC in the DIP package to the capacitor to less than 0.5 inches of power and 0.5 inches of ground path. This is a residual loop inductance from the decoupling capacitor to the IC pads of about 20 nH.

In contrast, in a printed circuit board, the minimum loop inductance can be reduced by almost a factor of 50 if done correctly. This is a significant advantage of using a printed circuit board. This is why it is so valuable using small-size MLCC capacitors with very short, wide leads with power traces over a ground plane from the capacitor to the IC pads.

To take full advantage of the PCB interconnects, the decoupling capacitors should be placed as close to the IC as possible with low mounting inductance. Anything else will increase the residual inductance and increase the switching noise.

Due to the longer paths in a solderless breadboard, there is a fundamental limit to how much the switching noise can be reduced. It is a good habit to always try to approach this limit. This means use the power and ground rail of the solderless breadboard to distribute power and add a decoupling capacitor with short leads between the two columns in close proximity to each IC on the solderless breadboard.

The value of the capacitor is not as important as its proximity, as long as it is larger than about 10 uF. A larger value, without sacrificing loop inductance, is better as it will also provide local charge storage for longer times and reduce the second source of switching noise from the IR drop of the VRM. **Figure 13.9** shows the placement of a 1000 uF capacitor between the power and ground rail of a solderless breadboard and the reduction in short-time switching noise with and without the capacitor.

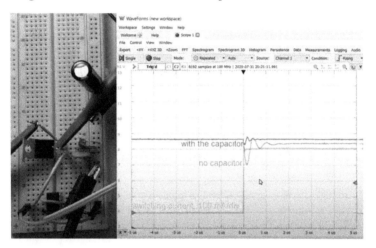

Figure 13.9 The switching noise measured on the collector of a transistor suddenly turning on 400 mA of current. The voltage drop noise is on a scale of 1 V/div. Without the capacitor, it is a 1.7 V drop. With the capacitor, it is 0.2 V drop.

The decoupling capacitor located about 1 inch away from the IC it is decoupling reduced the switching noise by almost a factor of 10, compared to without the capacitor. If it could be placed a little closer, it would reduce the switching noise even more.

Keep in mind that there are fundamental limits of how low you can get this switching noise using a solderless breadboard.

Watch this video and I will demonstrate how PDN noise is measured on a solderless breadboard and the role of decoupling capacitors.

13.6 The Power Delivery Path

In a solderless breadboard, the most important best design practices for the power path are to:

1. Route power to each device using the vertical columns in the solderless breadboard when possible. This reduces the resistance and provides an orderly distribution of power to multiple devices.

2. Add at least 10 uF capacitors in close proximity to the power pins of the IC on the breadboard. These connect between the power and ground connections with short leads.

3. When there is room on the solderless breadboard, add additional, large value electrolytic capacitors of at least 500 uF to provide local charge for longer periods of time.

4. Always use a voltage rating of the capacitors at least twice the DC voltage of the power rail. This is for reliability and to avoid capacitance reduction due to derating factors.

5. Always pay attention to the polarity of electrolytic capacitors and how they connect into the power rail. If the capacitor is connected in the reverse polarity, it may explode, as shown in **Figure 13.10**.

Figure 13.10 Why you should never connect an electrolytic capacitor in reverse polarity. It will explode and look like this.

An example of the placement of decoupling capacitors on a solderless breadboard is shown in **Figure 13.11**.

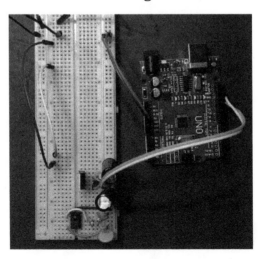

Figure 13.11 An example of a solderless breadboard with low inductance decoupling capacitors in the power and ground rail in close proximity to the ICs. They decouple the ICs from the result of the inductance in the power rail back to the VRM.

13.7 Inrush Current

The potential downside of using a large number of high-value capacitors on the power rail is the large inrush current when the external VRM is connected to the power rails of the solderless breadboard. An example of this situation is shown in **Figure 13.12**.

Figure 13.12 An example of a solderless breadboard with many large decoupling capacitors. Watch out for the large inrush current when power is applied.

When we connect the power rail of the solderless breadboard to the power source, the capacitors will need to be charged up by the VRM. This means there could be large initial currents flowing from the external power source into the capacitors. This is referred to as inrush current.

If the power supply has a source resistance like 0.5 ohms and the supply is 5 V, the initial inrush current that could flow from the power source to the capacitors could be as large as 5 V/0.5 ohms = 10 A. This is huge!

If the power is being supplied by a USB hub, the initial current load on your power rail could be enough to pull the voltage on the USB hub low enough to cause a fault on the USB hub of your device. This could cause a failure in your device with the USB hub.

Be aware of the danger of adding too much capacitance on your power rail and the potential inrush current surges it could create.

Of course, it lasts for only a very short period of time, roughly one RC time constant. If the capacitance on the rail is 1000 uF, the time constant would be RC = 0.5 ohms x 1000 uF = 500 usec. This is probably shorter than what would trip any fuses, but might trigger an error or warning.

An example of the inrush current measured on a circuit board with about 200 uF of onboard capacitance from a low output impedance USB hub source is shown in **Figure 13.13**.

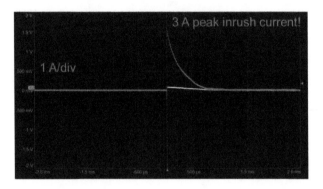

Figure 13.13 Measured current feeding the power rail of a board from a USB source. The peak current is 3 A but it only lasts for about 0.25 msec.

If the source impedance of the external power rail were higher, the time constant to raise the power rail on the board would be longer.

One fix is to add a soft start circuit using a series MOSFET which dynamically changes its series resistance based on the voltage on the rail. At startup when the rail voltage on the board is low, the MOSFET resistance is at some level to charge up the on-board rail. When the voltage has reached close to the final set value, the MOSFET's resistance is very low.

13.8 Summary of the Eight Habits for Using a SSB

1. Use the vertical columns for power and ground.

2. Use the two inner columns as ground the two outer columns as power or a consistent assignment.

3. Color code all the wires: red for power, black for ground.

4. Keep interconnect lengths short.

5. For long interconnect lengths keep the return wires connected to their signal wires.

6. Add at least a 10 uF capacitor adjacent to every power and ground connection to each IC with low loop inductance.

7. Keep all leads of all components as short as practical. Consider rotating resistors in a vertical orientation.

8. When practical, add an indictor LED to the power and ground rail to indicate when power is applied.

13.9 Practice Questions

1. What causes switching noise in the power path?

2. What are two geometry features that influence loop inductance?

3. What happens to the inductance of a loop if the current through it rises?

4. What is the typical minimum value of decoupling capacitor that should be used?

5. What does a decoupling capacitor decouple?

6. Once enough capacitance is provided in a capacitor, what is the most important quality of the capacitor?

7. Where should a decoupling capacitor be located?

8. What is an advantage of using a PCB over a solderless breadboard to reduce switching noise?

9. What is a downside of using too much decoupling capacitance?

10. What sets the ultimate limits to how low the switching noise can go?

Chapter 14
Design for Performance: The PDN on a PCB

The PDN consists of three elements, as shown in **Figure 14.1**:

- ✓ The power generator or voltage regulator module
- ✓ The power consumer or the active device
- ✓ The interconnects and components connected between the VRM and the active devices

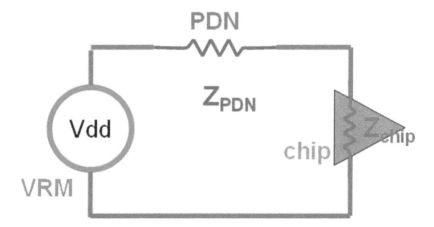

Figure 14.1 The three elements of the power distribution network. The resistor does not mean worry about the resistance of the PDN, but the impedance of the PDN.

The source of the voltage is the *voltage regulator module* (VRM). These are commonly an external AC to DC power supply, power from the USB connection, a battery, a switch mode power supply (SMPS), or a low drop out (LDO) regulator.

The consumers of the power are the *ICs* or other *active components*.

Connecting between the VRM and the power consumer devices is the *passive network of interconnects and passive elements such as capacitors.*

The combination of these three elements is the PDN.

Once connectivity is established, the only reason we will add a feature or do something special in the layout of a board is to *reduce an anticipated source of noise.*

In the PDN, there are four common sources of noise we want to manage:

1. The self-aggression noise from the VRM on its output.

2. The self-aggression noise from the IC switching currents on its own voltage rails.

3. The mutual-aggression noise from the IC switching currents on the rest of the board-level PDN.

4. The mutual-aggression noise from board power rail noise onto the IC power rail.

14.1 VRM specifications

Selecting the power source for your application has a number of considerations. The first question is mobility. Will it be:

- ✓ Plugged into an AC wall outlet?
- ✓ Plugged into a USB port?
- ✓ Battery powered?
- ✓ Using energy harvesting and a supper capacitor or rechargeable battery?

Once you decide the basic input method based on form and function, you need to determine the requirements for your design, including:

- ✓ The DC voltage required
- ✓ The tolerance needed
- ✓ The maximum transient current and for how long
- ✓ The average or typical current
- ✓ Your cost budget

Keep in mind that the higher performance you need, like the smaller required voltage tolerance or the higher the current, the more it will cost. Be prepared to trade off cost and performance.

14.2 Voltage Regulator Module

If we just add the voltage regulator module (VRM) to the board, we will find there may be excessive noise from the VRM. We refer to the noise the VRM generates by itself on its output under the typical load conditions as its self-aggression noise.

In the case of a SMPS, there is the switching noise of the pulse width modulation (PWM) of the gates to turn the higher input voltage source into a lower average output voltage. The switching frequency varies from VRM to VRM but is about 5 kHz to 500 kHz. To reduce the ripple of this switching noise we add a low-pass filter in the form of an LC circuit.

The inductor is usually an external element on the output switching node. It is on the order of 10 uH, based on the current load and the switching noise voltage you can tolerate on the rail.

A capacitor is added to turn the inductor into a 2-pole filter. We want a pole frequency well below the switching frequency so that the higher frequency components of the switching are filtered by the low-pass filter.

In a 2-pole filter, the transmitted signal is -40 dB down, or 1%, at a frequency 10x the pole frequency. This is why we try to select components to provide a pole frequency 1/10th the switching frequency.

We can estimate the amount of capacitance we need to add so that the pole frequency is well below the switching frequency. If we assume the switching frequency is 50 kHz, we want to engineer the pole frequency to be no higher than 5 kHz.

The pole frequency would be

$$f_{res} = \frac{1}{2\pi\sqrt{LC}} = \frac{159 \text{kHz}}{\sqrt{L[uH] \times C[uF]}} \quad (1.3)$$

If the inductor is 10 uH, and we want the resonant frequency to be 5 kHz, the capacitance we would add is at least 100 uF. This is where electrolytic or tantalum capacitors come in. We can easily get values in the 100 uF to 1000 uF range.

We also want to make sure the voltage rating of the capacitor is at least 2x the rail voltage to make sure we do not have a danger of the capacitor failing.

All electrolytic and tantalum capacitors are polarized. They have a + lead and a - lead. When assembling to the board, it is critical that this polarity be correct; otherwise the capacitor may explode.

Too large a capacitor may introduce a problem. Some SMPS have an over-current protection. When they turn on, if the capacitor value is too large, the inrush current to charge the capacitor may be large enough to activate the over-current protection. The SMPS may shut off, retry, and shut off again, going through oscillations. A soft start option on some SMPS decreases the inrush current.

In addition, if the source DC voltage is noisy, some of this noise may get through the regulator and appear at the output of the SMPS and look like self-aggression noise.

The power supply rejection ratio (PSRR) is a measure of how much of the noise at the input of the SMPS gets to the output. A PSRR of at least 40 dB will keep the noise at the output to less than 1% the noise at the input.

One way of reducing the voltage noise on the output of the VRM is to use a linear regulator as the last stage. These will typically have noise that is less than 5% the noise of a SMPS. The price you pay is they are not as efficient as a SMPS and may contribute to higher power dissipation in battery-powered applications.

In the case of an LDO, there is no switching noise, but there is a feedback loop. The output voltage is sensed compared to an internal reference voltage and a series MOSFET adjusted to change its resistance to keep the output voltage at a set level.

Sometimes, if there is a little noise on the output, and the output voltage can change rapidly, faster than the feedback loop, we can get oscillations in the feedback loop. This will appear as self-aggression noise of the LDO, which can be more than 1 V in amplitude.

The way to fix this problem is to add a capacitor at the output of the LDO to increase the time response for the output voltage to change. This prevents the oscillations.

The typical capacitor to add is from 1 uF to 50 uF, depending on the LDO. Too large a value and there may be too large an inrush current and the LDO may go through an auto shutoff. Too small a capacitor and there may still be oscillations.

A value of 10-22 uF is a convenient value to add at the output of the LDO as this is a value that will also be used elsewhere in the circuit for decoupling. When possible, the same parts should be reused to reduce the unique part count for the board. This will reduce the assembly costs and the risk.

The capacitor we add to the VRM to reduce switching noise or to minimize the self-aggression noise should be located in close proximity to the VRM. This is not a decoupling capacitor, it is a filter capacitor.

14.3　Self- and Mutual-Aggression Noise

After the self-aggression noise from the VRM is managed, there are two other important noise sources to control:

- ✓ The self-aggression noise of switching currents from a driver on its own on-die power rail.
- ✓ The mutual aggression noise from the board PDN noise onto very sensitive voltage rails.

The solutions to reduce these two problems are very different.

The fix for reducing the self-aggression noise from switching currents is to use low inductance decoupling capacitors in proximity to the power lead.

The fix for reducing mutual aggression noise is to add a filter from the onboard power rail to the noise-sensitive VCC pin of the IC.

14.4 Power and Ground Loop Inductance

We often refer to PCB interconnects as planar structures in that they live in the surface of the plane of the board. They are generally thin and wide and span some distance.

In a multilayer board, the loop inductance between the power path and a ground path on an adjacent layer is a little more complicated than just about the length of the traces and their spacing.

In planar structures with one wide trace over a plane, the loop inductance depends on three geometrical terms:

1. The *length* of the total signal-return path. The shorter the interconnect loop circumference, the lower the loop inductance.

2. The *separation* between the signal and return paths. The closer together the two conductors, the lower the loop inductance.

3. The *width* of the traces. The wider the traces, the lower the loop inductance.

14.4 Power and Ground Loop Inductance

Generally, the loop inductance is a complicated function of these three features and depends on the precise shape of the loop. In the special case of the trace width much larger than the dielectric thickness, as illustrated in **Figure 14.2**, the loop inductance is

$$L_{loop} = 32pH/mil \times h \times \frac{Len}{w}$$

where

L_{loop} = the loop inductance in pH

h = the dielectric thickness between the top and bottom conductors

Len = the length of the trace

w = the width of the trace

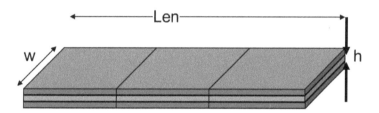

Figure 14.2 The dimensions of a wide strip on top of a plane when the aspect ratio of width to thickness is large.

This relationship points out that when the power trace is wide and close to the ground plane, the loop inductance will be decreased using a wider power path. This is an important additional feature to leverage for reduced loop inductance.

For example, if the power path is 200 mils wide and the dielectric thickness is 60 mils thick, a 1 inch long trace will have a loop inductance of about:

$$L_{loop} = 32pH/mil \times h \times \frac{Len}{w} = 32pH/mil \times 60 \times \frac{1}{0.2} = 0.38nH$$

Wide conductors close to a plane can have very low loop inductance. To reduce the loop inductance of the power-ground path, we use short lengths close together and as wide as practical.

When the trace width is not wide compared to the dielectric thickness, a rough estimate of the loop inductance of a trace and its return can be calculated using a variety of online calculators. In particular, the PCB Design tool from Saturn PCB, which can be downloaded to a PC, also has built-in estimators for the loop inductance.

For example, in the special case of a 6 mil wide trace 60 mils from a plane, a microstrip geometry, the Saturn PCB calculator estimates the loop inductance as about 20 nH/inch of trace length. This is the loop inductance including the power and ground path. The setup with this tool is shown in **Figure 14.3**.

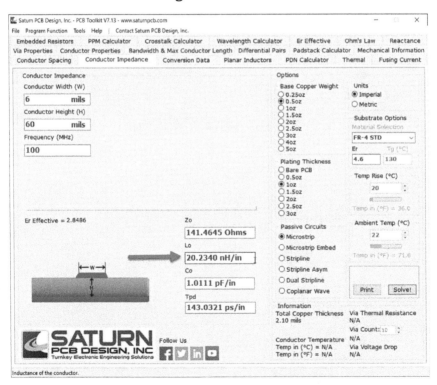

Figure 14.3 Using the Saturn PCB calculator to estimate the loop inductance per inch of the trace length, labeled as Lo.

A wider conductor, 12 mils wide, for example, would have a loop inductance of about 17.6 nH/inch.

14.5 Decoupling Capacitors

When an IC switches current to its outputs or just in its core logic, this current must ultimately come from the board level power rails, which then come from the VRM. This current can be switched through the IC very quickly from a few nanoseconds to a few microseconds.

Between the VRM and the pads of the IC is the inductance of the VRM. The transient current, flowing through this inductance, generates a voltage drop, the switching noise. This appears on the power rails of the IC. The way to reduce this noise is to reduce the loop inductance from the IC pads to the VRM.

As described in Chapter 13, a capacitor can provide the local charge storage during the switching time. These capacitors are called decoupling or bypass capacitors because they decouple the IC from the rest of the inductance of the PDN back to the VRM. They also bypass the transient current from flowing through the rest of the PDN.

The key design guideline for the decoupling capacitor is to reduce the loop inductance between the capacitor and the IC pads.

This means we want to use small-size, surface-mount capacitors, like MLCC capacitors. In printed circuit board applications when low switching noise is important, never use leaded capacitors, as these will have much more loop inductance in their leads than what we can achieve in a SMT MLCC capacitor.

In earlier chapters, we introduced three important best design practices of the power traces from the VRM to the IC. These design guidelines will reduce the path resistance and increase their current carrying capacity. In addition, we add the design guidelines for low loop inductance from the IC to the decoupling capacitors.

1. Use at least a 20 mil wide trace to route power.

2. Keep the resistance of traces delivering power acceptably low, so the worst-case IR drop is less than 3% the Vcc rail voltage. A 20 mil wide traces has 25 mohm/inch resistance. A trace 2 inches long will have a series resistance of 0.05 ohms. With 3 A, this will be 150 mV voltage drop, or about 3% Vcc.

3. Use a wide enough trace so it does not heat up with current. A 20 mil wide trace will handle 3 A.

4. Reduce the loop inductance from the IC to the nearest decoupling capacitor with short, wide connections for both the power and ground connections.

5. Use low loop inductance decoupling capacitors, such as MLCC capacitors. Never use a leaded capacitor for decoupling.

The minimum trace width for any power connection should be 20 mils. This can handle 3 A of current with no problems. And, more importantly, you can identify at a glance which traces are carrying power. They will be the wider traces.

> *There is no good compelling reason to use a copper fill for power distribution. It solves no problem and makes your board harder to debug.*

After the decoupling capacitors are routed to the IC pins, the second priority for routing should be the ground vias to the IC pins. This is to keep the loop inductance in this path as small as practical. These should be placed in as close a proximity to the IC as practical, routed with as low a loop inductance as practical.

The gnd vias on the other side of the capacitor should be routed with as short a connection to the bottom ground plane as practical.

There should be at least one 10-22 uF MLCC capacitor per power pin. If there is room, more capacitors can be added in parallel to reduce the loop inductance even more.

The longer the path length between the decoupling capacitor and the pads of the IC, the higher the loop inductance and the higher the switching noise. **Figure 14.4** is an example of the path from the IC to the decoupling capacitors in the same circuit, just with a different layout. Included in the figure is the measured voltage noise on the power rail of the IC when three I/O switch simultaneously. Lower loop inductance means lower switching noise.

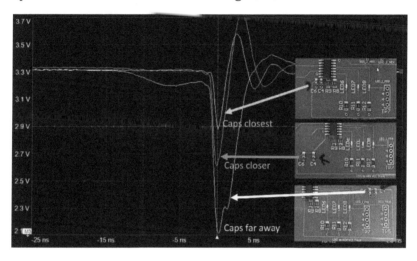

Figure 14.4 Measured switching noise on the power pin of an IC as the decoupling capacitor is moved farther away. To reduce the switching noise, keep the decoupling capacitor as close to the IC power pins as practical.

The preferred placement of decoupling capacitors is close to the IC they are decoupling and with a short, wide path, with multiple capacitors in parallel. Figure 14.5 shows an example of two different layouts for the same three decoupling capacitors. In each case, there are three 22 uF capacitors used. One design has high loop self-inductance, the other has reduced loop-self-inductance.

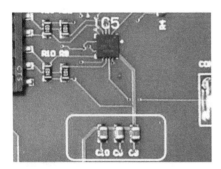 

Figure 14.5 Examples of the placement of the same three decoupling capacitors with good and bad layout. On the left is a larger loop inductance than on the right.

In the poor design, the capacitors are far from the IC, connected with a long, narrow trace. The return connections to the ground are short traces to vias, which is good.

In the better design, the capacitors are closer to the IC pads and have a wide path, a polygon path from the capacitors to the IC. This is referred to as a *power puddle*.

This design could be improved by adding a ground puddle at the ground side of the capacitors as well. It is the entire loop self-inductance of the entire power-ground path from the IC pads to the decoupling capacitors that we want to keep as low as practical.

14.6 A Decoupling Capacitor Myth; Part 1

Many application notes suggest using three decoupling capacitors for each IC, separated in value by a decade. A common recommendation is to use a 1 uF, a 0.1 uF, and a 0.01 uF capacitors, all in parallel.

One reasoning is that the smaller-value capacitors will have a self-resonant frequency at a higher frequency than the high-value capacitors and thus provide a lower impedance at high frequency and provide high-frequency decoupling.

This is completely wrong.

An ideal capacitor has an impedance that drops off like 1/f, forever. But a real capacitor has some loop self-inductance to the power and ground pads of the IC. This, plus its internal, equivalent series resistance means a real capacitor will behave like an ideal series RLC circuit. An example of the equivalent circuit of a real capacitor is shown in **Figure 14.6**.

Figure 14.6 Equivalent circuit model of a real capacitor as a series RLC circuit

This is a very good model for a real capacitor mounted to a circuit board. The C is the intrinsic capacitance of the capacitor. The R is the equivalent series resistance (ESR) due to the conductor plates in series with the capacitor. The L is the loop self-inductance between the MLCC capacitor and the pads on the IC. A comparison of the simulated impedance of an RLC model and the measured impedance of a real MLCC capacitors mounted to a circuit board is shown in **Figure 14.7**.

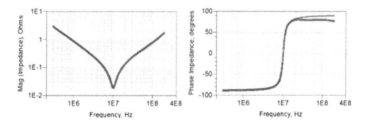

Figure 14.7 Comparing the measured impedance of a real decoupling capacitor mounted to a board and the simulated impedance of an ideal series RLC circuit with optimized values of R, L, and C.

The impedance profile of the real capacitor has a dip in the middle at the frequency referred to as the self-resonant frequency (SRF) where the reactance of the ideal C and L are equal but of opposite sign.

The self-resonant frequency is given by

$$f_{SRF} = \frac{1}{2\pi\sqrt{LC}} = \frac{159 \text{MHz}}{\sqrt{L[nH] \times C[nF]}}$$

The reasoning goes, if the equivalent series inductance, L, of each capacitor is about the same, the smaller value capacitor will have a self-resonant frequency higher and provide a dip, a lower impedance, at higher frequency. Put three capacitors in parallel, each with different self-resonant frequencies and there will be three dips in the impedance profile.

In this reasoning, the small-value capacitors provide lower impedance at higher frequency and act as high-frequency decoupling capacitors.

When three capacitors are added in parallel, such as shown in **Figure 14.8**, while there will be three dips, there will also be two peaks corresponding to the parallel resonances of adjacent L and C values.

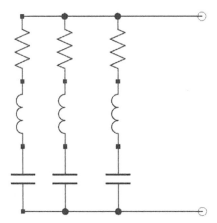

Figure 14.8 The equivalent circuit of three real decoupling capacitors in parallel.

The comparison of the impedance profile of two combinations of three MLCC capacitors, all with the same value of 10 uF, and three different capacitors with values of 10 uF, 1 uF, and 0.1 uF, are shown in **Figure 14.9**.

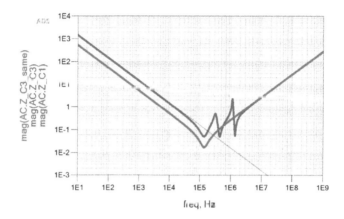

Figure 14.9 Impedance profiles of two combinations of three capacitors. Note for three different values, while there are three dips, there are also two peaks.

While there are three different self-resonance frequency *dips*, there are also two parallel resonant *peaks*. What generates the noise on the power rail is a high impedance. It is not about how *low* the impedance goes, it is about how *high* the impedance goes.

The parallel resonances create two peaks in the impedance profile. These peaks are what could potentially create more noise on the power rail than if all three capacitors were the same value.

It is the loop inductance of the multiple capacitors in parallel that causes the impedance to increase at higher frequency. This equivalent inductance is identical for the two combinations of three capacitors. They each have the same mounting inductance. At higher frequencies where the inductance dominates, their behaviors are identical.

Using three different value capacitors all with the same ESL offers zero advantage at high frequency but may create more noise due to the impedance peaks they create. In addition, they do not have as much capacitance at low frequency compared with three capacitors using the same value capacitance.

Do not follow the common advice, which is wrong, to use three different value capacitors.

14.7 A Decoupling Capacitor Myth; Part 2

The most important feature of a decoupling or bypass capacitor is its low loop inductance. When capacitors are through-hole technology with leads, it is generally correct that smaller-value capacitors come in smaller packages, and smaller physical size capacitors can be added to a board with a lower loop inductance. This lower loop inductance is good.

For example, a 47 uF capacitor is commonly available as an electrolytic capacitor. A 0.1 uF capacitor is usually available as a ceramic disc capacitor. And a 0.01 uF capacitor will generally be a smaller size than a 0.1 uF capacitor. Examples of these capacitors are shown in **Figure 14.10**.

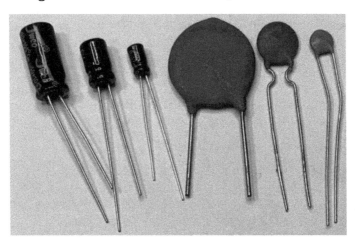

Figure 14.10 Examples of through-hole capacitors, electrolytic on the left and ceramic disc capacitors on the right. The smaller physical size capacitors can be mounted to the board with lower loop inductance.

The smaller the physical size of a leaded capacitor, the lower its mounted loop inductance. Between an electrolytic and a 0.01 uF small ceramic disc capacitor, the loop inductance might vary from 25 nH to 7 nH.

Since inductance dominates the impedance of a capacitor at higher frequency, it is quite correct to call the small-value capacitor with the smaller body and shorter leads a higher-frequency capacitor than the large-value capacitors.

This is why we sometimes refer to small-value capacitors on the order of 0.1 uF or even 0.01 uF capacitors as high-frequency capacitors, in that they will have lower impedance at high frequency.

This is fundamentally due to the structure of these leaded ceramic disc capacitors. It has nothing to do with their capacitance, it is about their physical size having lower inductance.

When we select capacitors as decoupling capacitors, we want a large capacitance for low-frequency performance and a low inductance for high-frequency performance.

This is the origin of the recommendation to add a 10 uF, 1 uF, and 0.1 uF capacitors to each power pin of an IC. It is assuming the capacitors are through-hole and in different body sizes and will have different mounting inductances.

When these capacitors are added in parallel, the impedance profiles at *low frequency* will look like the parallel combination of their capacitances and will be dominated by the larger capacitor.

At *high frequency*, their impedances will look like the parallel combination of their loop inductances, which will be dominated by the smaller-value inductance.

At intermediate frequencies, there will be parallel resonant peaks, the heights of which will depend on the amount of ESR of the capacitors. The higher ESR of electrolytical capacitors will generally damp out the parallel resonance peaks. An example of the individual impedance profiles of three leaded capacitors and their parallel combination is shown in **Figure 14.11**. There are no obvious parallel resonances because of the large ESR of the capacitors that damps out these parallel resonances.

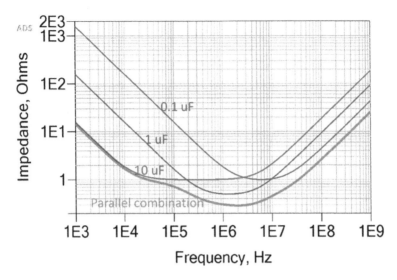

Figure 14.11 Impedance profile of three leaded capacitors and their parallel combination. Note the small-value capacitors have lower impedance at high frequency.

When you use through-hole capacitors, using different value capacitors for their different inductances is an important design guideline. This was a common design guideline 50 years ago when only through-hole capacitors were used.

However, with MLCC capacitors, it is possible to get 10 uF, 1 uF, and 0.1 uF capacitors all in the same body size, and all with the same loop inductance, and a much lower loop inductance than with a leaded capacitor. **Figure 14.12** shows examples of the same value capacitors in a leaded form factor and as an MLCC capacitor.

14.7 A Decoupling Capacitor Myth; Part 2

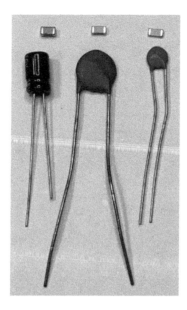

Figure 14.12 Examples of three different value capacitors as leaded and as MLCC. Each MLCC capacitor can have an identical mounted loop inductance.

Using MLCC capacitors, a 10 uF capacitor can have exactly the same mounting inductance as a 0.01 uF capacitor. For MLCC capacitors, all capacitors in the same body size are high-frequency capacitors.

When three different value MLCC capacitors are used, there will be parallel resonances with impedance peaks that may increase the rail noise. If the same body size MLC capacitors are used, they will all have exactly the same inductance and all be high-frequency capacitors.

A comparison of the impedance profiles for three legacy through-hole capacitors, three different value MLCC capacitors and three MLCC capacitors with the same capacitance are shown in **Figure 14.13**. There is no value in following the 50-year-old design guideline for leaded capacitors when using MLCC surface-mount capacitors.

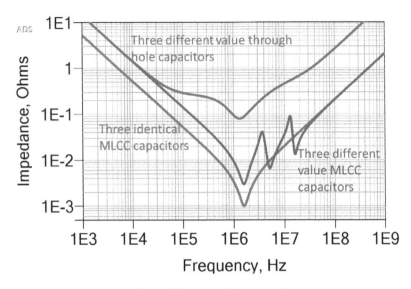

Figure 14.13 Comparing three different combinations of three capacitors: legacy through-hole, three different value MLCC capacitors, and three same value MLCC capacitors.

If you are going to use three different capacitors, the best alternative is to use all the same values. But, the only advantage of three individual capacitors is providing the lower loop inductance of their parallel combination. Generally, a single, low-inductance capacitor may be perfectly adequate.

Unfortunately, many designs created today carry over the legacy of designs created 50 years ago, before the widespread use of MLCC surface-mount capacitors. Many application notes still suggest using three different value capacitors, assuming the smaller-value ones have lower inductance.

When a single capacitor is selected for decoupling in a circuit, it is usually selected as a 0.1 uF or 0.01 uF capacitor, believing that a smaller-value capacitor will have lower inductance. **Figure 14.14** is an example of a design introduced in 2019, implemented in MLCC capacitors, yet retaining the older habits.

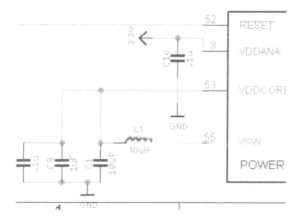

Figure 14.14 An example of a modern schematic with the legacy recommendation of using three different value capacitors. Note also the single capacitor on the 3.3 V line is 0.1 uF.

If you use these three different value capacitors, you will not have the lowest impedance possible at low frequency and will have parallel resonant peaks between the capacitors.

Read the full article about the myth of three capacitor values here.

14.8 Routing for Power Distribution

As part of the power distribution routing, the first step is to place the decoupling capacitors in close proximity to the power pins of the IC.

Then engineer the interconnects between the IC pins and the capacitor mounting pads with short, wide traces for the power connections as well as the ground connections to the bottom layer. When multiple capacitors are placed in proximity, a power puddle, which is a wide conductor to the multiple capacitors, can be used between the decoupling capacitors and the power pins of the IC.

The connections between device pins and gnd vias should be as short as possible to minimize the impact of blocking routing channels. Consider placing the ground vias under the IC footprint. Then the connections to the gnd via to each of the IC pins can be a short, direct path.

The goal is to reduce the loop self-inductance of all gnd or return path connections and the loop-mutual inductance of all signal-return path loops. This means NEVER share gnd vias. Every gnd path should have its own short path to the return path.

Once the local power distribution is routed, the connections from the VRM and the local decoupling capacitors can be routed with 20 mil wide traces.

Since the decoupling capacitors decouple the inductance of the PDN from the capacitor to the VRM, it is not necessary to take any special steps to reduce the loop inductance from the VRM to the capacitors. These traces can be routed as long traces to facilitate the routing. The design guideline is to use a minimum of 20 mil wide traces to distinguish power traces from signal traces and provide low enough resistance and high enough current-carrying capacity.

This is why a copper pour for power is not necessary. It solves no problem and makes it harder to debug the board routing.

Where appropriate, isolation switches and indictor LEDs can be added to the power distribution interconnects. The LED indicators on power rails will tell you at a glance that a rail is hot and has a voltage on it. The isolation switches will aid in debug to enable you to isolate circuits so others can be tested without interference.

Always be mindful of the maximum current flow in the power distribution. When it is likely to exceed 3 A, consider wider power traces.

14.9 Ferrite Beads

Many devices like microcontrollers will have multiple pins connected to power. The four most common types of power pins are:

- Power to the core logic, VDD
- Power to the I/O signals, VCC, or VIO
- Power to analog components, AVCC

- DC voltage to other specialized sections, like AVref

Generally, even though each of these pins might be connected to the same power rail net, at the very least they should have their own decoupling capacitor with as low a loop inductance as practical.

The current requirements for each power rail on the die will be different. The noise they generate on their power rails will be different. And their sensitivity to power rail noise will be different.

Digital power rails, like the VCCIO, VCC, or VDD rails, will generally have large switching currents. The problem to reduce for these power pins is the self-aggression noise from switching currents through inductance. This means use low-inductance decoupling capacitors.

For other sensitive rails, like a Vref or an analog VCC, AVCC, the current draw will generally be low, so the self-aggression noise may be small, but the pin may be extra sensitive to noise created on the power rail by other switching components.

When high-current devices switch, some of their self-aggression noise may get on the power rail on the board and pollute the power rail with their self-aggression noise. This noise may be seen on the power pins of other devices. This is a form of mutual-aggression noise: the noise on the PDN rail from other sources is what is picked up by other power pins sharing this rail and may interfere with the operation of the component.

Mutual-aggression noise has a completely different root cause from self-aggression noise and its solution is completely different. When mutual-aggression noise is the problem, you want to design the PDN path from the IC pin to the PDN with a *filter,* to filter out and reduce the noise from the board-level PDN getting on the sensitive VCC pin of the IC.

This filter is typically implemented as a simple low-pass LC filter by adding an inductor in series with a large-value, low-inductance decoupling capacitor. Adding some series resistance to the filter provides some damping.

These inductor components used to filter noise from the board-level rail from getting on the Vcc pin of a sensitive component are often called ferrites or ferrite beads because they are composed of a loop of wire wrapped around a core of ferrite ceramic.

The ferrite ceramic is an insulator with a high permeability. By adding this material to the core of the coil, the inductance of the coil is dramatically increased, sometimes by a factor of 100x. The ferrite material is also lossy, so it can attenuate higher-frequency currents. Damping resistance is a natural part of the ferrite inductor from the high resistance of the metallization and from the ferrite material.

This sort of ferrite is just an inductor with extra loss. The loss is an added benefit to damp out any resonances in the LC filter. This type of inductor is distinguished from a *power* inductor that is designed for high inductance but *low* loss. While a ferrite and a power inductor are both inductors, their intended applications are very different.

Examples of through-hole and SMT ferrites and their internal structure are shown in **Figure 14.15**.

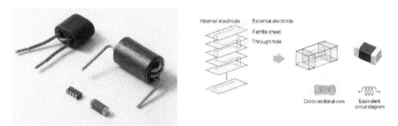

Figure 14.15 Left: examples of leaded and SMT ferrite filters, courtesy of Allied Components. Right: the internal structure of the coil in a SMT ferrite filter and its equivalent circuit and package style, courtesy of Murata.

To make the best low-pass filter to filter the noise on the power rail from getting on the power pin of the IC, the ferrite inductor should go between the board-level power rail and the AVCC pin of the IC. The decoupling capacitor connects directly to the pin on the IC and ground. This way, it acts as a 2-pole low-pass filter for any of the noise on the board to reach the pin that is sensitive to noise. **Figure 14.16** shows an example of the equivalent circuit.

14.9 Ferrite Beads

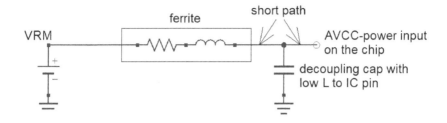

Figure 14.16 Circuit using a ferrite bead (an R and L element) in series with the decoupling capacitor.

To provide the lowest pole frequency so more noise is filtered out, the C should be chosen to be as large as practical but consistent with low mounting inductance. This is generally about 22 uF as an MLCC component.

The inductance of a ferrite is usually not specified explicitly, but the impedance at some frequency is. For example, the impedance may be specified as 2 kohms @ 30 MHz. If all of this impedance is due to the inductance, then the inductance is

$$L = \frac{Z}{2\pi f} = \frac{2k}{6.2 \times 30M} = 10 \text{uH}$$

When a 22 uF capacitor is used on the other side of the inductor, the pole frequency is

$$f_{pole} = \frac{1}{2\pi\sqrt{LC}} = \frac{0.159}{\sqrt{10 \text{ uH} \times 22 \text{ uF}}} = 10 \text{ kHz}$$

Generally, a series LC filter will have a peak in its transfer function at the pole frequency. The Q of this peak is

$$Q = \frac{1}{R}\sqrt{\frac{L}{C}}$$

To prevent ringing, we want the Q to be about 1. This means the series resistance should be about

$$R = \frac{1}{Q}\sqrt{\frac{L}{C}} = \frac{1}{1}\sqrt{\frac{10\ \text{uH}}{22\ \text{uF}}} = 0.67\ \Omega$$

The DC resistance of a ferrite inductor is usually specified in its spec sheet. For example, a typical value is 0.5 ohm series resistance.

The rated series resistance of 0.5 ohms is very close to the damping resistance needed to prevent a large peak in the transfer function.

Based on these values, we expect the transfer function of noise from the board level PDN to the power pin of the IC to show a pole frequency of 10 kHz, a very shallow peak and a -40 dB/decade drop-off above 10 kHz.

The model of the ferrite bead filter with the 22 uF capacitor on the IC power pin and the simulated transfer function is shown in **Figure 14.17**.

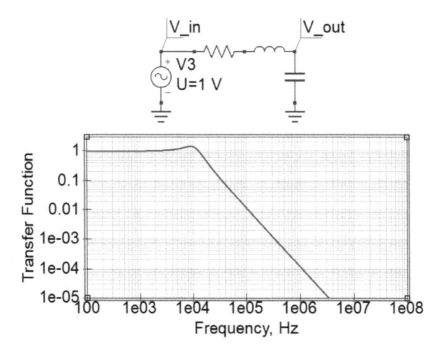

Figure 14.17 The equivalent circuit of the ferrite bead and decoupling capacitor and the transfer function from the board PDN to the IC power pin.

This will filter fast spikes from the power rail with frequency components above 10 kHz so they do not appear on the AVCC power pin.

The larger the inductance and larger the capacitance, the lower the pole frequency, and the better the filtering of noise from the board-level power rail to the sensitive power pin of the IC.

When the noise on the power rail is a pulse, for example, that lasts for 1 usec, its frequency components are above 1 MHz and will be filtered by the time it gets to the IC power pin. The simulated transient voltage on the IC pin from a 1 V spike on the board PDN is shown in **Figure 14.18**.

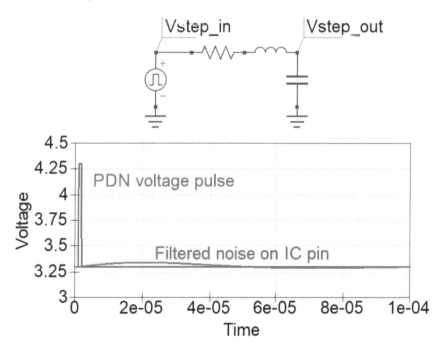

Figure 14.18 Simulated voltage noise on the IC pin filtered by the ferrite bead and decoupling capacitor from a 1 usec noise pulse on the board-level PDN.

This is why we use a ferrite bead in series with the board PDN to reduce the noise that would get on sensitive voltage pins.

If a ferrite bead works so well on filtering board-level PDN noise from the board on an IC pin, why not use a ferrite bead on all power pins?

There is always some DC resistance in a ferrite bead. The resistance is essential to damp out any peaks in the transfer function. This means there will be a DC voltage drop in the ferrite bead with DC current and less voltage getting to the device's IC power pin.

For a 0.5 ohm resistance, 1 A current draw in the PDN will result in 0.5 V drop to the IC pin. This may be too much IR drop.

Ferrite beads are suitable only on IC pins that do not draw more than about 100 mA and where the noise is generated on the board and must be filtered before it gets on the sensitive IC power pin. A ferrite will not reduce the noise on the IC power rail from self-aggression switching currents. The decoupling capacitor does this.

The worst design feature you could possibly do is add a ferrite to ground return path pins. This will dramatically increase the ground bounce noise on all signal-return paths that share this return pin. Yet this is a recommendation sometimes posted on online forums, such as shown in **Figure 14.19**.

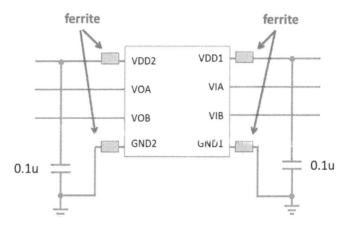

Figure 14.19 This is an example of a reference design found on the web. Adding ferrites on the ground return pins is an example of the absolute worst thing you can do in a design. This will dramatically increase ground bounce. This is an example of why you must take responsibility for your own designs and not believe everything you find on the internet.

14.10 Summary of the Best design practices

The best design practices for the power distribution path on a circuit board are:

1. Use a continuous ground plane on the bottom layer.

2. Use 1-3, 10-22 uF capacitors rated for 2x the rail voltage. Since a common rail voltage is 5 V rail, they should be rated for at least 10 V.

3. Place the decoupling capacitors as physically close to the package leads as practical.

4. Use as wide a trace from the IC leads to the decoupling capacitors as practical. This can be a small polygon copper fill with pads in the polygon. These are called power puddles.

5. Be sure to add thermal relief to the pads to make soldering easier.

6. Use as short and wide a path from the ground side of the capacitors to the vias to the ground plane as possible.

7. Never share a ground via among multiple decoupling caps. This will just increase the loop inductance. Always add individual ground vias, preferably using a short, wide trace from the capacitors to the vias. It could be a ground puddle.

8. Consider using the space under the IC to route the ground connections to the bottom plane. A short, wide path from the IC ground pad(s) to the bottom ground plane is as important as from the capacitor's ground side to the bottom ground plane.

9. Generally, always try to reduce the loop self-inductance from the IC leads to the capacitors, and their ground connections.

10. Never use a leaded capacitor for decoupling — it has 10x the loop inductance of a well-mounted MLCC capacitor.

11. Add at least a 22 uF capacitor to the input side of the VRM to reduce the noise at its input so it does not pass through the VRM.

12. Route power traces on the board from input to the VRM and from the VRM to the decoupling capacitors of each IC with 20 mil wide traces, providing the expected current is less than 3 A.

13. To prevent noise from the power rail getting on the sensitive voltage pins of an IC, filter it with a series ferrite inductor and a large-value, low-inductance capacitor on the sensitive power pin.

14.11 Practice Questions

1. Where should decoupling capacitors go?

2. What could you put under the IC footprint?

14.11 Practice Questions

3. What should you do for an assembly pad that will be placed in the middle of a plane?

4. Do you need thermal relief pads for signal vias?

5. Where should you place a test point you want to test with a 10x scope probe?

6. Does it matter how you route the power traces from the VRM to the decoupling capacitor? What is important about this path?

7. How much current can you put through a 20-mil wide trace before it raises 30 deg C?

8. How wide a trace do you need in 1 oz copper to carry 10 A of current?

9. What does a ferrite do?

10. What is the problem a ferrite filter solves?

11. How should you design a ferrite filter?

12. What are the three design features of a ferrite filter?

13. Why should you never put a ferrite on a ground pin?

Chapter 15
Risk Reduction: Design for Bring-Up

Of course, our goal in designing a product is to *get it right the first time*. This is where anticipating all the problems and designing them out is critically important.

But sometimes, even with best intentions, your product does not work the first time. The goal then becomes to *get it right the second time*. We need a strategy to find and fix the problems as quickly as possible.

15.1 Test is Too General a Term

Design for test (DFT) usually refers to testing the quality of a product in a production environment. This determines if the product passes some specific functional tests before it is considered to work and be shipped to the end customer.

DFT is about designing the board to make it easy to test for quality in a production environment. This means providing test points at strategic pads on the circuit board for test fixtures to make contact and perform automated tests. An example of a PCA being contacted by a bed of nails fixture connected to an automated tester is shown in **Figure 15.1**. This type of test is referred to as in-circuit test (ICT) and tests for basic functionality of the final board.

Figure 15.1 A PCA connected to a bed of nails fixture and then to an automated tester for quality testing in a production environment.

DFT depends a lot on the final test strategy of how you will test the quality of each board in production. It depends a lot on the specific tester, the fixture to connect to your PCA, and the automatic test equipment (ATE) you plan to use. The details of DFT are beyond the scope of this book.

In addition, there are other important tests all products must go through. While this book focuses on prototypes, it is important to put in perspective the range of tests all production-worthy products must go through at some point.

There are five different types of tests for a new product:

- ✓ Quality
- ✓ Margin
- ✓ Stress
- ✓ Reliability
- ✓ Compliance

Validating the *quality* of each board in production is about testing the assembled board for a manufacturing defect. This is accomplished by in-circuit testing of each part using automated test equipment and test vectors. This test answers the question, does this specific part meet specification right now today? DFT usually refers to this specific production level test each individual unit will go through.

Validating the *margins* of the product tests to see how much performance variation or noise there is compared to what is acceptable. If the margins are very small, a small perturbation in one component might cause the whole product to fail. When a million parts are manufactured, there will absolutely be variations in each part. If the margins are too tight, the percentage of units that can pass an initial test, the production yield, may be so low as to make the product too expensive to produce.

Measuring the margins under *stressed* conditions is an indication of how robust the product is under the extreme use conditions, over voltage range, temperature range and even user abuse. This is sometimes referred to as engineering validation testing (EVT).

This is done in the development phase of the product by measuring the performance of the product under extreme conditions. Does it still meet the performance requirements under extreme conditions, and will a high percentage of manufactured parts meet the performance requirements under the stressed condition?

Validating the *reliability* of the product verifies the product will still meet the performance specifications not just today, but over the specified lifetime of the product. Reliability testing is part of the design validation test (DVT) suite of tests.

While quality refers to the product meeting the expectations today, reliability refers to the product meeting the expectations over a long period of time. Everything wears out, everything ages, everything eventually breaks. How long the product lasts in the extreme conditions to which it may be exposed is a measure of its reliability.

If a product is expected to last 10 years, it is just not practical to turn on some number of units and wait to see if they are still working in 10 years.

Instead, you have to identify the potential failure modes, find conditions that accelerate these failures, and test at extreme acceleration factors, hoping you can extrapolate from the extreme conditions to the end-use conditions for the lifetime of the part.

This is the origin of highly accelerated life tests (HALT) and highly accelerated stress screen (HASS) testing. The most common failure modes are from

- ✓ Solder joint and metal fatigue failure accelerated by thermal cycling and thermal shock failures
- ✓ Mechanical joint failures due to high stress on joints accelerated by vibration and high gee force shocks, such as the drop test
- ✓ Corrosion-induced opens or shorts due to voltage, humidity, and temperature acceleration factors
- ✓ Wear-out failure in active devices accelerated by voltage, current, and temperature

The failure modes and acceleration factors are part of reliability engineering, which is a multidisciplinary field with extensive analysis.

Validating the *compliance* of the product to industry standards is a practical way of establishing some level of stress, margin, and reliability testing. It is often the only practical, cost-effective solution to quantify a very complex, hard-to-pin down set of behaviors.

There are a range of test conditions and criteria for passing created by the IPC, ASTM (American Standards of Testing and Manufacture, formed in 1898), the FCC, Underwriters Labs (UL), MILSPEC, and other industry organizations. Once the product passes a specific test, it is then *certified* as passing this standard.

For example, to be sold in the United States, a product must be certified to be compliant to the FCC part 15 class A or class B tests EMI. Many consumer products need to pass a UL test for safety. Any product used in military applications must pass a set of MIL standards tests.

Many of these tests are performed multiple times through the production phases, in the early prototype phase, in the early production runs, and in the final production runs. The earlier in the product life a problem can be identified and designed out or corrected by a part replacement or more robust design feature, the lower the total cost of ownership (TCOO) of the product from field failures and liability issues that might arise in the product.

15.2 What Does It Mean to "Work"?

Before any of these qualification tests are done, the very first step is validating the *design of the prototype*. This means verifying the selection of parts, the schematic design, the connectivity of the parts, and the physical design of the interconnects meets the performance spec established in the plan of record. This test answers the question, does your product work?

In a perfect world, you design your product, purchase the parts, assemble them together, turn it on and it just works, which means that it meets a level of initially defined performance or behavior specifications. You have validated the design and can move on to the other tests.

For example, if your prototype is supposed to measure the temperature of a sensor and flash a red LED if the temperature is too hot, a blue LED if the temperature is too cold, and a green LED if the temperature is acceptable, testing to this specification is straightforward.

You would turn the product on, put the sensor in a calibrated oven and test for the thresholds at which the LED colors change. If your product works to this specification, you are done with bring-up.

Move on to other tests or the rest of the product development cycle, such as cost reduction.

Part of the plan of record is to define what it means for the product to work. What are the basic functions that must be demonstrated to verify the prototype works?

For example, in an Arduino board, the board must

- ✓ Show a voltage of 5 V on the + 5 V power rail
- ✓ Show a voltage of 3.3 V on the +3 V power rail
- ✓ Appear as a device when connected to the USB port
- ✓ Allow uploading code from the Arduino IDE over USB
- ✓ Run a selection of test sketches like Blink

The process of testing to see if it works and if it does not, finding the hard errors and fixing them, if possible, to bring the product to an acceptable level of working is the *bring-up* phase. There are features we can add to the design to reduce the time spent in the bring-up phase. We call these features, design for bring-up (DFB).

When it does not work the first time, we enter the *debug* or *troubleshooting* phase to get it to work as quickly as possible the second time.

The most important step in troubleshooting your board to get it to work is finding the correct root cause(s) of the problem. Once you know the root cause, you can decide what you can fix in the current version of the product or consider re-designing the product to fix the other problems so the product meets the specifications.

Many products are complex. It may be difficult to debug a completely assembled product, especially if there are multiple problems.

There are some things we can do in the design of the product to make it easier and faster for us to find and fix the problems. These features we add may have nothing to do with the functioning of the product, and they may even add cost to the product, but they will save time and cost by shortening the time to find the problems and fix them. These features may reduce the risk of missing a schedule milestone. It is like buying insurance.

After the design is validated, some of the features added for bring-up can be removed as a cost reduction. DFB is an insurance policy to reduce the risk and time to get a working widget.

15.3 Design for Bring-Up

The most important step in fixing a problem is first identifying the *root cause* and having confidence you have the correct root cause. This process is referred to with various terms, all used interchangeably:

- ✓ Debugging
- ✓ Troubleshooting
- ✓ Forensic analysis
- ✓ Diagnosis
- ✓ Root cause analysis
- ✓ Bring-up

We use the same skills as a detective to find the guilty murderer, or a doctor to diagnosis an illness, or an auto mechanic to debug the source of the rattle.

A problem or error introduced to the design that will cause the product to fail is a hard error and must be fixed to validate the design. A problem which will not affect the functioning of the product but may add more noise or time in assembly or risk of making a mistake but not interfere with the bring-up is a soft error.

Connecting an enable pin low when it should be connected high is a hard error. Connecting the RX of one device to the RX of the other device instead of to its TX is a hard error. Using a 22k resistor instead of a 22 ohm terminating resistor is probably a hard error.

Using the wrong footprint for a part is a hard error. Making a 3 mil gap between a signal line and a pad may be a hard error if it results in a short in your board.

Hopefully, many of these hard errors can be fixed in the bring-up phase to get your product to work. They can be implemented as a redesign in the next spin of the hardware.

Forgetting to place the pin 1 identifier in the footprint of a part is a soft error. You can still get the board to work if you pay attention to how the part is placed on the board. This soft error increases the risk and the assembly time, but your board will not fail if you are careful. Not putting the name of the layout file on your board to track it will not affect the functioning of the board, but will increase the risk of not using the correct file for assembly. It is a soft error.

Sometimes, if you don't have the time to make all the changes, it may be acceptable to release a design with known soft errors. The reduced time to get your board back and assembled to validate the design may be worth the increased risk from soft errors.

Once you have validated your design works, you can use the next design spin to implement the changes to correct the soft errors that may reduce noise, reduce risk, reduce assembly time, and increase value in the user.

Sometimes the changes we have to make to fix an error are in the design: the component selections, the schematic, or the layout. Sometimes the changes we have to make are in the assembly of a part to the board. Sometimes the changes we have to make are to replace a defective or damaged part. Sometimes it is a software change.

Some problems can be fixed quickly. If a lead is soldered poorly, identified by visual inspection with a microscope, touching up the solder joint can fix the problem in minutes. If it is a single wire change, sometimes a little surgery on the board of cutting a trace and adding an engineering change wire can fix the problem. An example of a board with an engineering change wire is shown in **Figure 15.2**. These are added to fix either a design problem or a manufacturing problem.

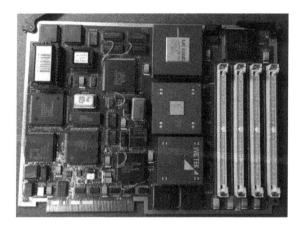

Figure 15.2 Examples of engineering change wires on a prototype computer board to correct design problems.

Engineering change wires are sometimes referred to as white wire changes or green wire changes, depending on the standard color wire a company or team uses for its rewiring.

But if the fix requires a new component or a new layout, the time to get a new board back and assembled may delay the product by 5-10 days, and then new problems may be introduced. This is why it is so important to find all potential problems as early in the design phase as possible.

The general principle behind the design, bring-up, and troubleshoot phase, is

> *The earlier in the design cycle you can find and fix or eliminate a problem, the lower the final cost in dollars and time.*

15.4 Add Design for Bring-Up Features

If the product does not work the first time, the key step before we can fix the problem is to find the root cause. When designing a prototype, a risk reduction step is to assume the product may not work the first time and implement features that will help you get the product to work the second time as quickly as possible. These are features for bring-up, debug, troubleshoot, and forensic analysis.

There are five specific features you can add to your initial product to make it easier to debug the first prototypes:

1. Partition the product into functional sections and isolate either the power delivery or signals with isolation jumper switches.

2. Use indicator LEDs for power rails and critical signals.

3. Add test points to facilitate connecting test instruments like scopes and scope probes to power and signal nodes (including return path connections!).

4. Add either 3-way switches or dummy 1206 0 ohm jumper resistors to provide on-the-fly options for pull-up or pull-down or alternative component selection that can be manually implemented.

5. Consider adding small-value resistors in the series path of the power delivery as current sense resistors to measure the current draw of various circuits. A 0.1 ohm resistor will generate a 10 mV signal with 0.1 A of current.

The time to think about adding these features to make it easier to bring-up your product is at the very beginning when you are planning the product design.

The features to include in the BOM, schematic, and layout that will facilitate bring up, troubleshoot, and test may add component count, cost, and board real estate. This is part of the cost of buying insurance.

However, what you buy for this higher initial cost is the possibility of a shorter time to bring up your board. The purpose of adding bring-up or troubleshooting features is to reduce the time to get it right the second time.

15.4 Add Design for Bring-Up Features   357

Once the design is found to be working and meeting the performance requirements, some of the bring-up features can be removed from the production board to reduce costs or the size of the board.

You should get in the habit of adding bring-up components and features as part of the BOM and as additional elements in your schematic.

It is much harder to debug a board after you have assembled all the parts and you power it on for the first time. If you are lucky, everything works. *But luck should not be part of your design strategy.*

If you are assembling your board yourself, do not go ahead and assemble all the parts to the board all at one time. Assemble and test in stages. This way you can isolate any problems as you go.

If you are using an assembly shop to assemble your board, you do not have the option of assembling your board and bringing it up as you go.

Instead, your product should be partitioned into isolated circuits that can be turned off and separately tested after the board is assembled. *This will make it easier to find the root cause of errors.*

This means you should add jumper switches to isolate the power delivery or specific signals to different functional parts of your circuit.

Turn off all the sections you can and test each block to verify it works as expected. Then successively turn on a new section and verify it works as expected. A simple 2- or 3-position pin header with shorting flags is a low-cost switch. An example of three of these used in a board is shown in **Figure 15.3**.

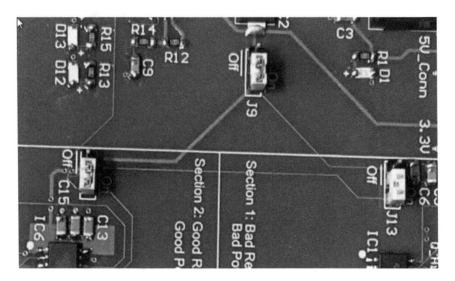

Figure 15.3 An example of a board with three header pin switches used to turn off sections of the board.

After you have verified the complete design is correct and all the parts have been selected correctly, you can remove some of the features implemented just to facilitate the bring-up phase to reduce the board cost or board size. If these features — isolation switches, LED indictors, and test points, do not impact cost or performance, you can keep them in the final design.

15.5 Jumper Switches

Jumper switches can be used to isolate power to specific circuits or signals in the signal flow. Just think through the impact of the added loop inductance of a header pin and a flag. A simple jumper switch will have an added series loop inductance of about 10 nH.

Never add jumper switches between the IC and its decoupling capacitors, as this path is very sensitive to loop inductance, with the goal to reduce it as much as possible.

Jumper switches can be as simple as 100 mil centered header pins placed in series with circuit elements, such as shown in **Figure 15.4**.

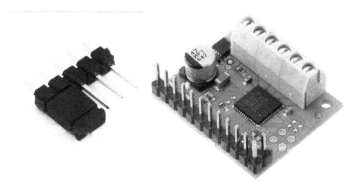

Figure 15.4 Examples of header pins and shorting flags. Use either 3-pin header pins to shift the flag over or add a row of no-connect header pins to store the flags when not in use.

When a two-pin switch is used, a shorting flag is placed over the pins to make the shorted connection. But when the switch is meant to be open, where do you place the shorting flag? If it is left off the board, there is a high risk it will be lost.

One possibility is to rotate the shorting flag 90 degrees and stick the flag on only one pin. The advantage of this approach is it uses just a 2-pin header and you can tell at a glance the open and closed connection setting. The disadvantage is that the flag is not held on very securely and can easily fall off and get lost.

If you want your circuit to connect to two different inputs or outputs, you can use a 3-pin header pin. The center pin connects to the device and the two pins connect to the two options. When the shorting flag is placed in one position, it connects one option. When it connects in the other position, it connects the other option to the device.

Using header pins and shorting flags is a low-cost, simple way of adding options to your prototype board.

15.6 LED indicators

An indicator LED should use the minimum current necessary to still show an on or off state so it does not consume much power.

This is generally about 1 mA. With a 3.3 V rail, a 2 V drop across an LED, the voltage drop across the current limiting resistor of any LED circuit is 1.2 V. Using a 1k ohm resistor would result in about 1 mA of current.

Unless there is a strong compelling reason otherwise, use a 1k ohm resistor as current limiting in all LED indicator circuits. Using a 1k resistor with all LED indicators will also keep the unique part count down.

Use LED indicators on all power rails so that at a glance you can tell each power rail is energized.

Use LED indicators on all important digital I/O such as communication busses or critical signals.

The LED indicator circuit should be in parallel with a signal line, never in series. (Why not? What is the problem created by placing them in series?)

Figure 15.5 shows a board with examples of LED indicators, test points, and switches to isolate parts of a circuit.

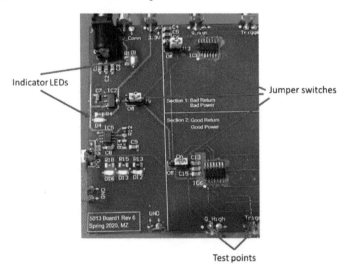

Figure 15.5 An example of a board with features to facilitate test and bring-up including test points, jumper switches, and LED indicator lights.

15.7 Test Points

The purpose of a test point during the bring-up phase is to facilitate measuring a voltage at a specific node or terminal. This will be either a signal voltage or a power rail voltage.

Here are some important guidelines to consider when adding test points:

- ✓ When using a test point for a scope, be sure to add a ground reference point in proximity.
- ✓ The design of a test point is based on how you will probe the board.
- ✓ Be sure to label each test point — the signal AND the gnd connection. Just labeling it as test point, or signal, says nothing. Even TP2 is vague. Consider TP2_QuietHI, for example. A poorly labeled test point that makes the tester work to figure out to what it connects is a soft error to be avoided.
- ✓ It is just as important to label the ground connection of the test point so that you can tell instantly how to make the connection to the scope.
- ✓ You must have a test plan BEFORE you complete the schematic.
- ✓ Include test points and how they will be probed in the schematic.
- ✓ Estimate signal bandwidths compared to measurement bandwidths and interconnect bandwidths.

Here are some options to use for basic test points:

- ✓ Small hole near the board edge for a micrograbber clip or 10x probe clip. Diameter should be 60 mils and located with its center about 60 mils from the edge of the board. This is shown in **Figure 15.6**.

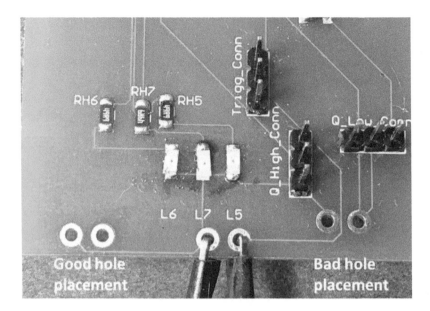

Figure 15.6 Example of two test points, as plated through-holes, 60 mils drill diameter with a center-to-center pitch of 120 mils. Their centers are 60 mils from the edge of the board. Note the holes on the right are too far from the edge of the board to clip onto, and that none of the test points have good silk screen labels.

- ✓ A large, exposed pad near the edge of the board to clip a ground lead.
- ✓ A plated through-hole to solder a wire to connect a gnd clip. It could be for an insertable test point or just a small hole to solder a wire into. Does it matter where the gnd clip is located?
- ✓ A scope probe's 10x probe with two via holes spaced 300 mils apart, 40 mil drill diameter.

The 10x probe with the ground spring tip has many advantages to other probing methods:

- ✓ It uses a low-cost footprint on the board
- ✓ It can have a measurement bandwidth in excess of 200 MHz
- ✓ It has low rf pick up

15.7 Test Points   363

✓ The 10x probe can be free-standing and allow for hands-free probing

Examples of a scope's 10x probe using a spring ground lead with two simple test probe footprints and the 10x probe inserted into the circuit boards, free standing, are shown in **Figure 15.7**.

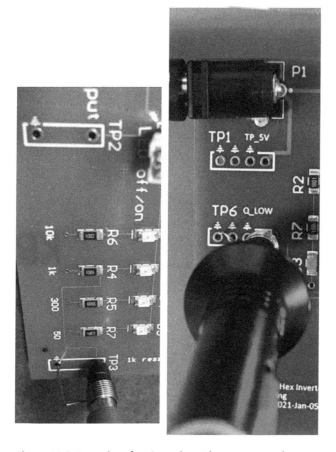

Figure 15.7 Examples of a 10x probe with spring ground tip inserted into a test point on two different boards designed for the 10x probe.

The 10x probe test point is probably the most common test point to use. The cap of the 10x probe is removed and the small spring clip that comes with all 10x probes is wrapped onto the tip. **Figure 15.8** shows a 10x probe tip with the cap removed and the spring ground tip attached compared with the conventional cap on and its ground wire.

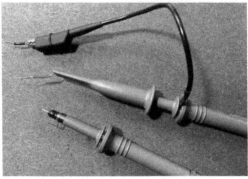

Figure 15.8 Example of a 10x probe with the spring tip ready for assembly (top) and comparing the signal-return loops of the spring tip and conventional wire ground wire (bottom).

When the probe with its spring tip is inserted into the plated holes in a circuit board, it is free-standing and hands-free. Multiple probes can be inserted into the board if the spacing between each test point is far enough apart.

There are two sizes for 10x probes: tips that are 2.5 mm in diameter and tips that are 5 mm in diameter. Obviously, the ground spring is different for each 10x probe, so it is important to be aware of which type of 10x probe you use. But the tips are both on 300 mils centers. Generally, the 5 mm tip is more robust than the 2.5 mm tip and is preferred. They can be used with exactly the same test points.

The test pattern for the 10x probe tip is two holes, 40 mils drilled diameter, spaced with 300 mil centers. A modification of this test point is to use three ground holes spaced on 100 mil centers. This is free, takes up no additional board area and allows for standard 10x probes and high bandwidth probes that may use 100 mil centers between their signal and return pins.

For the highest bandwidth connections and the lowest rf pickup noise from the near field of the board or the environment, a coaxial connection to the board should be used.

There are two commonly used coax connections to a circuit board: the SMA connector and the U.FL coax connector. Their main value is to provide a high bandwidth connection to the board to connect to a scope and to minimize the possibility of picking up near-field emissions from the board or other test points.

Examples of coax connections are shown in **Figure 15.9**. These are optional components to add to the board in order to pick up test signals using a coaxial connection, which should be less susceptible to rf pick up and less distortion from the loop inductance of the tip.

Figure 15.9 Examples of the two coaxial test fixtures. Left: SMA type connectors, edge-mounted and through-hole. Right: U.fl connectors, mounted to a circuit board, courtesy Sparkfun.

15.8 The Power Rail as a Diagnostic

The very first things a doctor does when checking your general health is listen to your heart, and measure your pulse rate and your blood pressure.

The functioning of your heart is an important diagnostic of your general health. If it is off, it is a general indication there is something else wrong with you. But, even it is normal, it does not mean you don't have another problem.

Likewise, the voltage noise on the power rail and the current draw to specific devices is an important diagnostic for the general health of a circuit.

You should include test points on each power rail to monitor the voltage with a scope to search for any noise synchronous with specific functions of your circuit. In addition, the current draw by the whole circuit or specific components is also a useful diagnostic.

There are three general methods of measuring the current draw of the power rail:

- ✓ An external current clamp scope current probe
- ✓ An integrated Hall effect isolated current probe
- ✓ A series resistor acting as a sense resistor

An external clamp on current probe measures current by measuring the magnetic field around the wire. The probe clamps on the outside of the wire. Magnetic fields in the wire circulate inside the probe, which contains a high permeability material. Inside this high permeability material is a Hall effect sensor that measures the magnetic field.

After calibration, the magnetic field measured by the Hall effect sensor is a direct measure of the current through the probe. An example of a probe is shown in **Figure 15.10**.

15.8 The Power Rail as a Diagnostic 367

Figure 15.10 An external current clamp Hall effect sensor measuring the current in a circuit, courtesy of Teledyne LeCroy.

The external clamp current sensors can have a very high bandwidth, at least as high as 100 MHz. This means a scope can measure current fluctuations synchronous with some clocks switching.

To use this sort of probe, you must add a wire loop about 1 inch in diameter somewhere in the current path, to slip the clamp around. This can be implemented by two plated through-holes in the series path of the current to a device. An external wire is added for testing and then the two holes are connected with a short loop after bring-up is complete.

A new family of current sensors based on the hall effect can be integrated directly onto your board. The current to sense passes through a low-resistance shunt path. Inside the sensor is a Hall effect probe that measures the local magnetic field of the shunt conductor. This is a direct measure of the current in the shunt. An example of the internal structure of this class of insitu current sensing probes is shown in **Figure 15.11**.

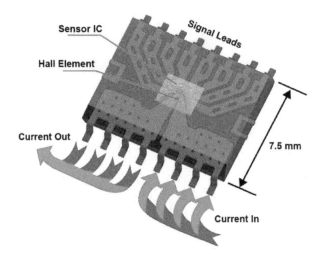

Figure 15.11 Internal structure of a shunt current sensor based on the Hall effect, courtesy of Allegro Microsystems.

The advantage of this method is that it has a minimal impact on the current in the power path.

The third and simplest and lowest cost approach is to add a low-value series resistor in the power path and measure the voltage across it. The voltage across the resistor is a direct measure of the current through it. These resistors are referred to as current sense resistors.

Using a sense resistor is very simple to use. The value of the resistor should be high enough to create a large enough voltage to be easily measured, but small enough so the voltage drop created is small enough in the application to be insignificant.

If the voltage across the resistor can be kept below 100 mV, this is generally not an issue in most power rail applications. If the current is about 100 mA, the sense resistor would be R = V/I = 100 mV/100 mA = 1 ohms. A 0.5 ohm resistor is a typical sense resistor value to use for these low-current applications.

15.8 The Power Rail as a Diagnostic

An example of a 0.5 ohm sense resistor on a circuit board is shown in **Figure 15.12**. The resistor is in the series path from the voltage source on the board to the rest of the power rail distributed on the board. By measuring the voltage across this sense resistor, the dynamic current draw of the entire board can be measured in real time as the circuit operates.

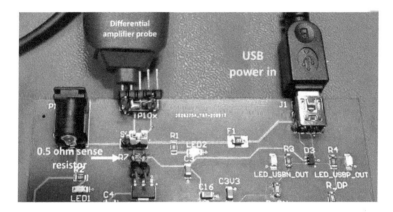

Figure 15.12 Sense resistor mounted on a circuit board with sense lines to a connector for a differential amplifier connected to a scope.

In this application, a differential amplifier is required to measure the voltage across the resistor. One side will be at the VCC and the other side will be at VCC − 0.05 V. This could be measured with an external differential amplifier or even an on-board instrumentation amplifier.

For example, it is possible to monitor the current draw from an I/O pin of a component on the board at the same time as the current draw to the power rail on the board.

All the fast-changing current from the I/O is supplied by the local decoupling capacitor adjacent to the microcontroller. This allows the current through the VCC pins to turn on as quickly as the driver can switch. However, the current into the board will be much slower, as it is just to charge up the decoupling capacitors. **Figure 15.13** shows these two very different transient responses. The current into the board changes much more slowly than the current from the I/O.

370 *Practical Guide to Prototype Breadboard and PCB Design*

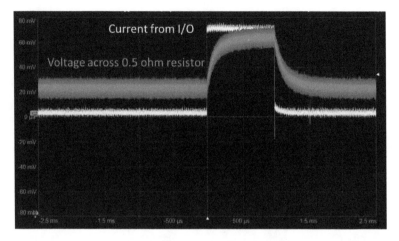

Figure 15.13 The measured current from an I/O using a sense resistor on the output of the I/O and the current from the power rail charging up the local decoupling capacitor with a much slower transient response.

Another important application of a sense resistor is to measure the transient current from the power source when the board is initially plugged into power. There will usually be a large inrush of current used to charge up all the on-board decoupling capacitors connected to the power rail. This initial current when the board is first powered-up is referred to as the inrush current.

An example of the inrush current into an Arduino board with about 200 uF of total decoupling capacitance, which flows from the USB power source, is shown in **Figure 15.14**. The sense resistor is 0.5 ohms. A voltage across the resistor of 1 V means there is 1 V/ 0.5 ohms = 2 A of current. The measured inrush current from the USB port peaks at 2 A of current!

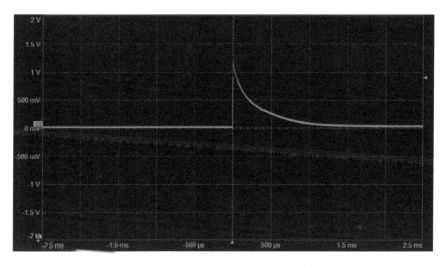

Figure 15.14 Measured voltage across the 0.5 ohm sense resistor on board showing a peak inrush current of 2 A when the Arduino board was plugged into a USB port.

15.9 Practice Questions

1. What does design for test refer to?

2. What is the difference between quality and reliability?

3. What is an acceleration factor?

4. When is it necessary to think about how you will test your prototype?

5. What are three features you could add to your board design to make it easier to bring-up and test?

6. When designing a test point for a scope, what should you consider?

7. If you expect to see an inrush current of 1 A, what is a reasonable sense resistor to add in series to the power path?

8. Why should you NOT use a 10x probe connected across a sense resistor in the series path of the power rail?

9. What is the best way of measuring the voltage across a sense resistor?

10. Why is the current flow into the PDN from the external power source a different signature than the current flow into the Vcc pin of a switching device?

Chapter 16
Risk Reduction: Design Reviews

As a general principle, the earlier in the design cycle you can find and fix an error, the lower the cost and schedule impact.

How much time is spent on reviewing a design, compared to actually working on the design and implementing it, is based on the balance between the importance of risk assessment and mitigation, and the cost in resources in time, money, and personnel to spend on review or on implementation.

If the design is man-critical, a lot of resources spent on design reviews may be worth the additional costs to reduce the risk of failure.

At the very least, you should implement two pauses in the design flow for two formal reviews, the PDR and the CDR.

16.1 The Preliminary Design Review

The PDR should happen early enough in the design cycle so that fundamental changes can be made in the direction of the design without seriously impacting the schedule.

Usually, the PDR is implemented after the plan of record is drafted. It is a review of the proposal.

Will the functions as proposed meet the performance goals? Even if the product works as intended, is the intention the correct intention?

A widget is designed to measure the temperature and display it on an LCD display. This may be a great function, but if the requirement is to also log the temperature on a website, it misses an important feature.

Are the important risk sites identified and a mitigation plan in place?

If your product uses an Atmega 328 microcontroller as a critical component, have you checked the delivery time? If the delivery time is 6 weeks, is there another part you can choose or another place you can find this part?

Sometimes the PDR is be done after the schematic is complete so it can be included in the design review. This is still before any significant costs are incurred.

16.2 The Critical Design Review

The CDR should be done before significant funds are expended. This should happen before expensive components are purchased or the fab is ordered. You should schedule the CDR after the layout is complete, but before the fab and major components are ordered.

The schematic review can be included in either the PDR or the CDR depending on the schedule. Or, more commonly, a separate schematic design review is scheduled. This allows for a chance to correct flaws in the schematic before they are propagated to the layout.

Most PCB design tools have an electrical rule checker (ERC) that checks the schematic for some obvious design rule violations, such as unconnected pins or two different nets connected together, or a net that has only one pin connected to it. While this will find some obvious errors, the ERC tool has no idea of the intent for your design. It cannot find the error if the enable pin is supposed to be high and you have it connected low.

After an ERC, the author of the schematic should look it over and scan for obvious errors. Using a color code to the nets will often help errors stand out.

A schematic says nothing about signal integrity or power integrity or EMI. The wires used to establish connectivity are ideal transparent interconnects with no impedance and no delay. Looking at a schematic you can say nothing about the expected noise or how to potentially reduce it.

In this respect, signal integrity and power integrity and EMI live in the wires and the white space of the schematic. The best you can do in a schematic is place associated parts in proximity, mark some nets as critical nets either by color, width, or name, and add some notes on the schematic page as a hint to the layout.

In advanced tools, some nets can be categorized into net classes. A net class defines some conditions to a net for its routing based on the type of signals that might be carried by the net. A net class might be for controlled impedance transmission lines, or length matched differential pairs, or traces that are very sensitive to the routing topology.

Each net class may have a different set of layout constraints, such as line width for current handling or for impedance control. The more layout information we can list in the schematic or assign to a net class, the less chance errors may crop up in the layout.

The default net class for signal traces should define a 6 mil wide trace with a continuous return plane, and 13 mil drilled via holes for signal paths, and 20 mil wide traces for power paths.

After you think the schematic is complete and you have reviewed it and performed an ERC, it is time to have it reviewed by another pair of eyes. A fresh pair of eyes can often see a feature that you have stared at for hours. This is also a good test of the primary guideline for creating schematics: it should be easy to read and follow the circuit function.

One of the important principles to follow in searching for possible errors is the saying that an expert is someone who has made all the mistakes possible.

If you make an error, remember it and DO NOT make it again in a future design. Keep a list of the errors you have made, or you have seen others make, to use as a checklist for all future designs.

There are three important tricks you can use to help you search for errors:

- ✓ Look for obvious problems that stand out.
- ✓ Look for errors you have made before.
- ✓ Look for errors listed in a checklist.

Here is a checklist of questions to answer to avoid commonly made errors in the design and the schematic:

1. Is there at least one decoupling capacitors per power pin, all about 10-22 uF? In the schematic, are they located in close proximity of the device they are decoupling, just as a reminder for the layout?

2. Are associated parts in close proximity to each other in the schematic? Are decoupling capacitors near the IC they decouple? Are components in the power delivery path nearby each other, and test points, switches, and LEDs close to the components to which they connect?

3. Are all net connections to all pins connected? Are any important nets labeled with a new name so that the net name appears in the layout while you are routing?

4. Does each component have a footprint that is matched to the datasheet? Use the footprint manager to verify footprints to each component in the schematic.

5. Are all resistor values used for LED indictor lights the correct value, typically 1k unless otherwise specified?

6. Are all LEDs connected in the correct polarity?

7. Is the TX from one device connected to the RX of another device and vice versa?

8. Is the D+ line routed to the correct input pin and the D- line routed to the other input pin?

9. Are there any unconnected inputs that might take on arbitrary values due to ESD events?

10. Have you reviewed the datasheets of all specialized components and verified the part you specified in your BOM is the part in the datasheet?

11. Is the worst-case total power draw from your components below a reasonable limit for your power supply?

12. Do the power rail voltage requirements for each component match your power distribution?

13. Do crystals have 22 pF capacitors on either side to gnd? The 1 meg resistor across the crystal is optional if the inverter is guaranteed to oscillate.

14. Are all the enable pins set to either high or low, depending on the datasheet requirement?

15. Are there specific part value components you can eliminate to reduce the number of unique parts?

16. Have you included isolation jumpers where appropriate?

17. Have you included LED indicators on all important voltage rails?

18. Have you added tests points based on a specific probing method?

19. Is your schematic clear and easy for someone to read?

20. Are you using a good balance between net names and direct wire connections?

21. Have you added any clarifying text on the schematic page that will help you remember abbreviations or layout instructions?

16.3 DRC for DFM in the CDR

A design rule checker (DRC) is a built-in tool that checks the layout for obvious violations based on the rules established in the layout setup, such as metal features too narrow, soldermask slivers too narrow, or silk screen too close to a pad edge.

The DFM checkers that will check your design have no idea what the intent of your design is. All they can do is check for specific and commonly observed layout problems, that would violate specific manufacturing rules.

A good EDA PCB layout tool that has the manufacturing design rules built in can be set up to only allow layout features to be added to your design if they do not violate any design rule. This is called correct by design. The acceptable features in the layout tool are usually set up in the constraint manager in the layout tool.

In more advanced designs that are driven by signal integrity, another level of design rule checker can be used that checks the features for specific signal integrity performance conditions.

16.4 DRC for Signal Integrity

Signal integrity design rules are only important in high-speed boards where signal integrity problems dominate the circuit performance.

As a default condition for all traces that are not controlled impedance, the following signal integrity features should be implemented regardless of whether the interconnects are transparent or not:

- ✓ A low-inductance decoupling capacitor should be located in close proximity to the IC it is decoupling.
- ✓ All signal lines are routed with 6 mil wide traces and 13 mil drill diameter vias.
- ✓ All power nets are routed with 20 mil wide traces.
- ✓ All traces from a pad to a ground via are short.

- ✓ Ground vias are never shared. This just increases the potential ground bounce noise.
- ✓ Avoid a copper pour on a signal layer. It solves no problem and can contribute noise.
- ✓ A continuous return plane is routed on the bottom layer.
- ✓ A minimum number of cross-unders are routed in the ground plane.
- ✓ Any cross-unders in the ground plane are short.
- ✓ When a cross-under cannot be made shorter than ½ inch, a ground strap is added to provide a return path.

These features can benefit ALL designs. In addition, there are other design features to implement when the rise times decrease into the realm where interconnects dominate performance.

An example of a list of DRC rules that can be integrated into an automated layout checker tool to test for common signal integrity problems is in this summary article for example.

16.5 Layout Review

When implementing the routing of traces in the board layout phase, there are risks of introducing problems due to inattention.

Of course, during the layout process you will pay attention and complete all the routing and silk screen labeling correctly. But sometimes, errors slip in.

The first risk reduction step is to review what you have done before you release the design to fab. You should maintain a list of common errors to check for and each time you complete a board design project and find another hard or soft error, add it to your list.

Here is a starting-place list of the most common errors to avoid. Use this as a checklist for completion of the layout before releasing to fab. Add to it based on each design cycle:

1. Are all the ghost (rats nest) traces routed?
2. Are all the signal traces 6 mil wide?
3. Are all the power traces at least 20 mils wide?
4. Are all signal vias 13 mil drill diameter?
5. Are all vias created as NOT thermal relief unless there is a strong compelling reason to use thermal relief?
6. Are decoupling capacitors placed close to the IC with low loop inductance?
7. Are return paths continuous?
8. Are all cross-unders as short as practical?
9. Are all the ground connections from pads to the ground plane short?
10. Are all the switches, LED indicators, and test points included and located in the correct places?
11. Are the labels in the silk screen for the switches, LED indictors, and test points easy to interpret without looking them up on a table?
12. Are all parts labeled with part designators in close proximity to the part?
13. Are polarity indictors or pin 1 designators placed in the silk screen on the outside of where the part will be placed so it can be seen with the part assembled?
14. Are all the pads of one part associated together to identify where the part goes down?
15. Are the silk screen labels large enough to be legible?
16. Are any silk screen markings on top of solder pads?

17. Are all the test points labeled correctly with the associated signal names?

18. Is there any copper pour on signal layers?

After the DRC has been run and reviewed and the layout creator has done a thorough review, the last step in the risk roduction is to apply another pair of eyes to review your design. This is the peer design review process.

The final risk reduction step is the CDR, which is a more formal design review process for your team and the person who will actually pay for your board. By the time of the CDR, all the obvious errors should have been caught and fixed.

16.6 Practice Questions

1. What is the difference between a PDR and CDR?

2. What are two examples of potential problems that should be caught in a PDR?

3. What are two examples of potential problems that should be caught in a CDR?

4. What does an ERC check for?

5. What does a DRC check for?

6. Why is a PDR and CDR important to complete?

7. Why is it always useful to have another pair of eyes review a schematic or layout?

8. In which review should the availability of parts be explored?

9. Why is it a good practice to have a checklist of items to review in a design review?

Chapter 17
Step 2: Surface-Mount or Through-Hole Parts

There are a gazillion different part sizes and types of packages, and all have different names that are usually abbreviated with TLAs (three-letter-acronyms) or FLAs (four-letter-acronyms). This makes selecting parts really confusing. Get used to it.

All of these myriad parts can be grouped into a few general classes of part and body styles. When the same part is available in different body styles, it is very important to know what each package style is and understand the good and bad trade-offs for each body style.

17.1 Through-Hole and Surface-Mount

At the very top level, there are generally two types of parts defined by how they are assembled to the board. There are *through-hole* parts and there are *surface-mount* parts. These two types of parts are illustrated in **Figure 17.1**.

Through-hole parts   Surface-mount devices

Figure 17.1 Examples of through-hole and surface-mount parts.

A through-hole part has leads and gets inserted into holes in a board, then soldered to pads surrounding the holes. Surface-mount devices (SMD), also called surface-mount technology (SMT) parts, require no holes in the circuit board. They just get soldered to pads on the surface of the board.

The only thing holding a SMT part to the board is the adhesion of the copper pad to the laminate. In manual soldering assembly or repair, if the pad is heated too hot with a soldering iron, the pad may delaminate. Any pressure applied to the part and it may pull the pad off the board. This is why it is important to limit the temperature extremes when manually soldering parts to a board.

Through-hole is an older technology, while SMT is relatively newer. What drives the use of SMD parts is miniaturization. The smaller the product form factor, the smaller the SMD parts needed. Cell phones and other consumer products drive the use of ever smaller parts. They all use SMT parts. **Figure 17.2** shows an example of a small form-factor board covered with SMD components.

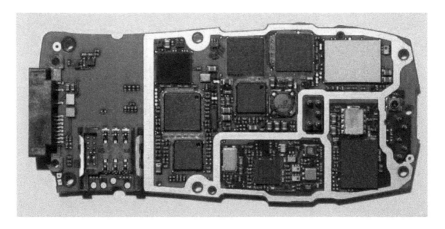

Figure 17.2 Example of a cell phone circuit board with many SMD devices on the small board.

If you are doing a hobby project and cost, size, and production are not important, using all through-hole parts might be an option.

But if you expect to produce boards by automated assembly, or make many copies of the board, or want to use the latest ICs, SMD parts will always result in a lower-cost board, smaller size, and when it matters, capable of higher performance with lower noise.

There are many IC components, such as microcontrollers, that are only available as SMD.

Once you make the decision to use SMD, consider using ALL of your parts as SMD. This will make the assembly process, part procurement, and unique part inventory easier and lower cost.

There is one exception to using a SMD. If any part will experience mechanical stress, such as a connector, consider using a through-hole version.

> *In a board with all surface-mount parts, the one place you should consider using a through-hole part is in a connector that might experience mechanical stress. The stress on a surface-mount part might pull the pads off the board.*

Because of their many advantages, get used to surface-mount parts in your design.

17.2   Types of SMT Parts

There are generally two classes of SMT parts: two-terminal and multiterminal. The terminal is the connection on the ends that get soldered. **Figure 17.3** shows examples of these two types of parts.

SURFACE MOUNT PACKAGE SIZES

Package type	Size in inches	Size in mm
0201	0.024" × 0.012"	0.6 mm × 0.3 mm
0402	0.04" × 0.02"	1.0 mm × 0.5 mm
0603	0.063" × 0.031"	1.6 mm × 0.8 mm
0805	0.08" × 0.05"	2.0 mm × 1.25 mm
1206	0.126" × 0.063"	3.2 mm × 1.6 mm
1210	0.12" × 0.10"	3.2 mm × 2.6 mm
2020	0.20" × 0.20"	5.08 mm × 5.08 mm
2512	0.25" × 0.12"	6.35 mm × 3.0 mm

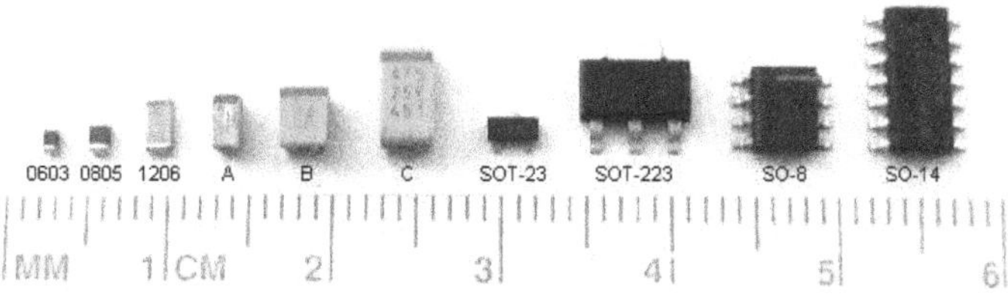

Figure 17.3 Examples of two-terminal and multiterminal type SMD parts.

Two-terminal parts are like resistors, capacitors, and LEDs. We label their specific part types in terms of their dimensions. The first number in the name is the distance between the terminals. The second number in the label is across the width. For example, a part that is 120 mils between terminals and 60 mils wide is referred to as a 1206, pronounced twelve-oh-six part. We leave off the last 0 in its dimension.

The common part sizes are 1206, 0805, 0603, 0402, and 0201. In many cell phones there are parts as small as 01005. A pepper shaker is used to place them on a board (*just kidding!*).

There is a limit to how small a part can be handled and assembled to a board. At some point in your career, you should experiment with assembling different-size parts to find the limit to what you can do if necessary and then routinely.

Generally, the larger the part size, the easier to assemble. With a little practice and the right techniques, anyone can routinely assembly 1206 parts to a board. A level of skill is required to manually assemble 0805 or even 0603 parts.

17.2 Types of SMT Parts 387

Using automated pick-and-place assembly equipment, 0402 parts can universally be assembled in volume. Because these size parts are so commonly used, generally, for the same value component, the 0402 body size will be the lowest cost. This is why some commodity part values are lower cost than others. If they are used a lot, there are more produced, and their price will be less.

An example of a small desktop automated pick-and-place machine is shown in **Figure 17.4**.

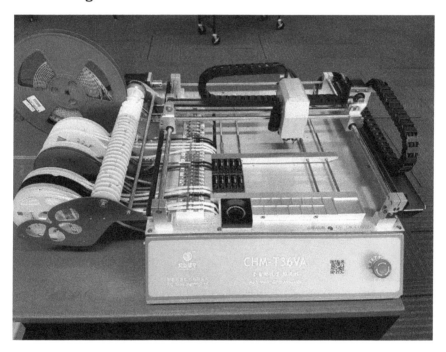

Figure 17.4 An example of a desktop pick-and-place machine that picks up a selected part from the reels on the left, aligns it with the camera in the lower middle, and places it on a circuit board in the center.

As a general best design practice, you want to use the smallest-size part you are comfortable assembling for the process you intend to use. Using a larger-size part, even if it is a little more expensive, may be an acceptable risk mitigation step to avoid assembly problems, if you want the option to manually assemble or repair the part.

If you plan on doing manual assembly, unless you have a strong compelling reason otherwise, plan on using 1206 part sizes for all of your projects. This is the lowest risk path.

If you plan to use an assembly shop, generally using 0402 parts is acceptable since all automated assembly tools can handle 0402 parts and they are generally lower cost.

17.3 Integrated Circuit Components

Multiterminal parts are usually integrated circuits (ICs). There are a gazillion different package sizes and styles. They generally fall into two classes, those with leads sticking out which make them easy to manually solder, and those without leads. These just have pads underneath the part. **Figure 17.5** shows examples of packages with leads sticking out and with no leads.

Figure 17.5 Examples of leaded and no-lead packages.

Generally, the package will have a smaller footprint if it has no leads, but it will be harder to assemble by hand.

Inside each of these packages is the actual IC that does all the active work on a board. An example of the IC exposed after the plastic is dissolved away is shown in **Figure 17.6**.

Figure 17.6. Example of the IC inside a plastic package with the plastic dissolved away. This is what is inside every plastic package. Source of the image is SparkFun.

Using a part with no leads, and only pads underneath it, like a QFN (quad flat no-lead) package or a BGA (ball grid array) part is a large risk site if you assemble by hand. These parts are very difficult to place on the board with their terminals underneath the package aligned to the pads on the board, and can only be reflow soldered. *If you plan to assemble your boards by hand, avoid using these parts.* It is only practical to assemble these parts with a pick-and-place machine and a reflow oven or hot plate.

Within the category of SMD, leaded part, there is an alphabet soup of names. **Figure 17.7** shows a few examples. A great source of pictures of packages and their names is the Topline website.

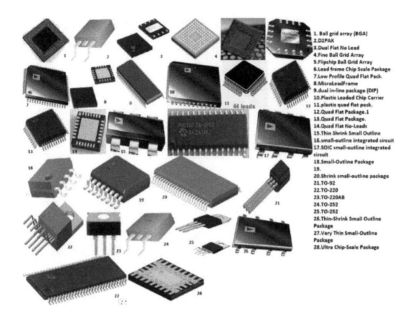

Figure 17.7 Examples of the wide mix of IC package types

Each letter in the name of a package style refers to some quality:

Q is quad, meaning leads are on all four sides of the part.

P is plastic as the molding component. This means non-hermetic.

C is ceramic, which usually means hermetic.

D is dual, referring to leads on two sides.

F is flat.

BGA is ball grid array.

CSP is a chip-scale package with a size not much larger than the IC.

WLP is wafer-level package and is bare silicon chip with some plastic.

N is no leads. This is usually just exposed pads.

SO is small outline, meaning a smaller package.

T is transistor package or thin.

IC is integrated circuit.

V is very.

For example, an SOIC package is a small-outline IC package. This is usually with dual leads on a 50 mil pitch.

It is not necessarily the case that one package style is significantly better than another. While they differ in the number of leads, the lead type, and the pitch, generally, their design is driven by lower cost in high-volume manufacturing.

One of the reasons there are so many different types of packages is that each package style is patented by an assembly or package vendor. To get around the patent of a competitor, they make a small change and call it a different type of package.

The key feature to pay attention to in selecting a multiterminal package is the lead pitch. Generally, a lead pitch, the spacing between the centers of adjacent leads, of 50 mils (1.25 mm) is easier to manually place, align, and solder to the board than a package with a lead pitch of 25 mils (0.6 mm). But they are larger packages for the same number of leads.

If you are manually assembling your board and you have a choice, select parts with a larger lead pitch. And to reduce risk, practice assembling parts with a lead pitch of 25 mils. Many ICs will only come in a package with a lead pitch of 25 mils. If you are using automated assembly, each vendor will specify the smallest body size part they can assemble and whether they can do BGA assembly. **Figure 17.8** shows some examples of package names, body style, and lead pitches.

SOIC
SMALL OUTLINE
Integrated Circuit

DRAWING	NOMENCLATURE	BODY WIDTH	LEAD TYPE
 8 - 16 PIN	SO = Small Outline	156 mil	
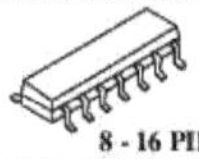 8 - 16 PIN	SOM = Medium Outline	220 mil*	Gull 50 mil Pitch
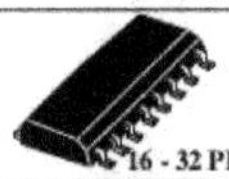 16 - 32 PIN	SOL = "Large" Outline SOP = "Small" Outline Package	300 mil	
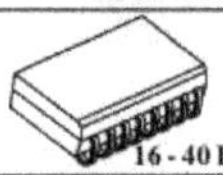 16 - 40 PIN	SOJ or SOL-J = "J" - Lead Large Outline	300 mil*	J-Lead 50 mil
 32 - 56 PIN	VSOP = Very Small Outline Package	300 mil	Gull Wing 25 mil
 8 - 30 PIN	SSOP = Shrink Small Outline Package	208 mil	Gull Wing 25 mil
 20 - 56 PIN	QSOP = Quarter Small Outline Package	156 mil	Gull Wing 25 mil

*Up to 440 mils

Figure 17.8 Examples of IC packages with lead pitches identified.

Read this article about SMD and through-hole parts.

Read this article about various SMD part labels.

This is a reference article from Topline about part names.

A good review article about IC packages and what is inside from Sparkfun.

17.4 Practice Questions

1. If a SMD part is lower cost than a through-hole part, why would anyone ever want to use a through-hole part?

2. In a product, what sort of parts should always be chosen as through-hole?

3. What keeps a surface-mount part from falling off the board?

4. Why should you avoid QFN or BGA parts when you are hand assembling?

5. If you are assembling a SMD resistor part by hand, what is the preferred size?

6. In an 0603 part, what are the dimensions of the part?

7. What are three advantages of using SMD parts over through-hole parts in a board?

8. What does a lead pitch of 50 mils refer to?

9. Which is easier to assemble manually to a board, a 25 mil pitch leaded package or a 50 mil pitch leaded package?

10. What is the difference between a DIP and an SOIC part?

Chapter 18
Finding the One Part in a Million

There are two families of parts you will use: commodity parts and specialized parts. Commodity parts are the commonly used parts that are easy to find, with many options such as resistors, capacitors, LEDs, and regulator ICs.

The noncommodity parts are specialized and may have a limited number of vendors. These parts may be harder to find and more care should be taken in identifying the specific part you need, where it can be purchased, and the delivery lead time. These are the specialized microcontrollers, op-amps, and sensors, for example.

As part of your risk assessment, you will need to identify the noncommodity, specialized parts you will use and determine if you can get them in the time frame you need, with the performance and features you require, and at a cost you can afford.

The noncommodity parts will make up the core of your BOM.

Part names can be very confusing. Here is a very good article that puts the naming of parts in perspective.

Pick parts that are popular. This will give you a higher chance they will be available when you come to order them.

18.1 An Important Selection Process

Whenever possible, try to reuse a part you have used before and use it multiple times in the same circuit. This is preferred over selecting new parts you have never used before and each component being a different value on your board.

A part you have already used means you have experience with it and there is less chance of being surprised. It is closer to being qualified for use in your applications. You have confidence the symbol and footprint information are accurate.

Do you really need to use three different op-amp part numbers or can you use the same op-amp for each circuit? Do you really need a 10 uF capacitor and a 22 uF capacitor or can you use two 22 uF capacitors or two 10 uF capacitors in all your applications?

Reducing the number of unique parts can reduce risk and cost. Sometimes it is worth it if the performance impact is minimal. For example, when selecting an oscillator frequency, if the exact value is not important, select a value you can achieve with existing parts in inventory rather than selecting new values.

When assembling a board with automated pick and place, there is a limit to the number of different reels the machine can use. When the unique part count exceeds the number of reels available the assembly cost will increase.

Some small assembly machines can handle 20 different reels. If the unique part count can be kept below 20, it may reduce the assembly cost.

Each different part added to your design introduces a risk point. If the part is not qualified, will it meet the performance spec you expect? There is the risk of the unknown. Can you use a previously used part instead of a new part?

The fewer the unique parts, the less inventory of parts you need to keep around. The process of reducing the number of unique parts in a BOM is called *scrubbing* the BOM.

If you can scrub a BOM count from 70 to 50 unique parts, there may be a price break in assembly.

The fewer unique parts means less chance of confusing one part with another, or placing the wrong part on a pad. Small SMD resistors and capacitors often do not have markings. They are easy to confuse.

Reducing the number of different capacitor values means less chance of confusing them.

Wherever possible, consider consolidating the number of unique parts and always ask if a new value is really necessary, or a more common and previously used value can be used.

This is the essence of engineering: design trade-offs. In selecting specific parts, you have to balance all the options and the resulting performance at what cost. The better you understand your application and circuit, the better you will be able to perform best cost-performance-risk trade-offs.

Keep in mind there is often more than one right decision. Just be aware of why you make the decisions you do.

18.2 Trade-offs in Selecting Parts

As you construct the details of your schematic in the third step, you will encounter the ordinary, common, or commodity parts, such as resistors, capacitors, LEDs, and header pins. Many ICs or other active devices are so commonly used they are commodity parts, such 74xx digital parts, 555 timers, op-amps, and NPN transistors.

When selecting each part, you will constantly be trading off:

- ✓ The performance
- ✓ The package style
- ✓ The delivery time
- ✓ The special features
- ✓ Previous experience
- ✓ The values of specific specs
- ✓ The cost

For example, a 555 like this one is $0.38, and operates at a maximum of 100 kHz. But, this one, operating at up to 3 MHz, is

$1.10. Is it worth an additional $0.72 for 3 MHz operation? If the intent is to run at 10 kHz, the additional price may not be worth it.

But, if a future design might need 3 MHz operation, it may be worth it to use the higher-price part in the low-performance application, so that you can gain experience with it, and keep just one 555 timer part in inventory. After you evaluate it in your initial application, it can become a qualified part and used for all future projects. This is part of risk reduction.

Even something as simple as a 1206, 51 ohm resistor, 5% tolerance, can vary from $0.028 to $0.28 each for a total of 10 parts, for nominally the same sort of part. Which is a better part?

When selecting parts, we will constantly be making trade-offs between performance, features, tolerance, size, delivery time, vendors, and price.

When there are many options to choose from, how do we find that one-in-the million best part to use?

18.3 The Search Order to Select a Part

The very first step when selecting a specific part is to decide on the assembly approach: through-hole or surface-mount.

The first step in selecting an IC is to decide the minimum acceptable performance. What feature in it is most important? If you are selecting an op-amp, what specs are most important: the input offset voltage, the input bias current, the gain bandwidth product, the DC drift, or the input noise voltage?

Then decide on the package style based on manual assembly or automated assembly. If manual assembly, what is the smallest size you are comfortable with? Unless you are skilled in the art, use 1206 as your smallest part.

Search on a distributor's website, such as Digikey, for the part and use criteria such as:

18.3 The Search Order to Select a Part

- ✓ Active (it is currently available)
- ✓ Type
- ✓ Mounting type
- ✓ Package style

The first-round search is to reduce the number of options to less than 500.

Do not be overly constraining initially.

For example, in searching for a 555 timer, just using these criteria of SMT and 8 pin SOIC reduced the options from > 500 to less than 125.

Make note of the various performance differences that influence the price and decide it if is worth it to use a more expensive but higher performance part.

If it is your intent to purchase just a few items, then type quantity of 1 in the View Prices at: box.

Sort the options based on Unit Price by clicking the up-pointing arrow — this will sort the smallest price at the top of the list. An example of the list is shown in **Figure 18.1**.

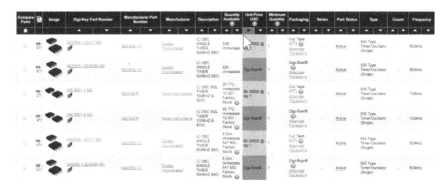

Figure 18.1 An example of searching for the right 555 timer.

The lowest price is around $0.37 for a 500 kHz part.

When you have found the right balance of performance and feature, narrow the search to find the last few candidates.

As a final step, always double check the minimum quantity and the lead time.

Keep in mind that often there is more than one right part option from which to select. Once you find one that is suitable, you will gain confidence in it and potentially reuse it in other designs.

Add this part to your favorites and reuse it in future designs. If you are already familiar with a part, it reduces an important risk of using a new part and finding out either your judgment is wrong about what the part does, if the part is actually the wrong size for your application, or the datasheet is misleading.

Here is a general algorithm to select parts:

1. Select the general category of the part you need.

2. Select the package style preferred. This generally means the smallest part size you can easily assemble, such as 1206. If you expect to use production assembly, most assembly shops can routinely assemble 0402 parts.

3. Select in-stock, datasheet, as these will be useful when integrating into your design flow.

4. If you have no strong compelling reason to select a specific value in a category, such as vendor or application, leave it alone.

5. Pick a range for the values you could use. For example, if a resistor, select a range of ±10% about the value you need if the value is not critical. Some specific values of capacitors are cheaper than others because they are ordered by large manufacturers in the gazillions.

6. Generally, component values in multiples of 10, 22, or 47 are lower cost because they are produced more. A resistor value of 47 ohms is more common than 50 Ohms. A capacitor value of 22 uF is often lower cost than a 35 uF capacitor.

7. If the tolerance is not important, select a value of 1% and higher. Generally, a 1% tolerance SMD resistor costs the same as a 5% tolerance. Check the pricing and do not order a tighter tolerance than you are willing to pay for. Generally, SMT parts are manufactured with tighter tolerances with smaller cost adders compared to through-hole parts.

8. When the number of items in the selection is less than 200, you can stop the filter criteria and look at the parts in the list.

9. Select the option for viewing 500 items per page. This makes scrolling through them so much easier.

10. If you selected the in-stock checkbox, all the parts in the list should be available now. But you should verify this in the parts list column.

11. When the list of parts is reduced to less than about 200, search the list by

    ✓ Unit cost.

    ✓ Available in stock.

    ✓ Minimum order volume of 1, or select the number you plan to buy in the quantity box. This will adjust the price to do a direct comparison.

    ✓ Trade-off part value with cost. Maybe use a lower voltage or larger tolerance for lower cost, as long as it is acceptable.

    ✓ Keep in mind there are often multiple right answers.

    ✓ Try to select a reliable vendor, if you can tell.

12. If you select a part you have not used before, you run the risk that:

    ✓ The vendor is unreliable and will not deliver.

    ✓ The part will come in and its footprint will not match the expectation.

    ✓ The performance of the part will not meet its datasheet.

> *This is why it is a good habit to reuse a part you have experience with.*

13. Sort by lowest price first. The two arrows at the top of each column are not really arrows. They are pyramids, showing small at the tip to large at the base or large at the top and small at the base.

14. Scroll down the list and look for the lowest priced part available in the smallest quantity you can purchase. Then look at the other choices, either value or part size, to choose the balance you want between performance, cost, risk, and schedule.

15. You should end up with 2 or 3 different items from which to select your specific trade-off balance.

16. Make sure your parts are available now.

17. Make sure the minimum quantity is within your range.

18.4 Selecting Resistors

When selecting resistors or capacitors, there are a few special criteria to use. Usually, surface-mount resistors and capacitors in 0402, 0603, 0805, and 1206 body sizes are called chip resistors or chip capacitors.

The tolerance. This is the absolute accuracy of the resistance, comparing the actual resistance to what a calibrated ohmmeter would measure. Unless you need a tight tolerance for an application, use a tolerance no tighter than 1%. For SMD resistors, there is little difference in the price of a 10% tolerance or 1% tolerance. A tighter tolerance will generally be more expensive. Use 1% or even 5% unless a more accurate value is important or does not cost extra.

18.4 Selecting Resistors

The value. Initially, select a range of values. Price is often related to the volume sold. Some values may be cheaper because that value part is made in larger volume.

If the specific resistor value is not important, select a range first. In some circuits, we use a resistor as a current-limiting element and the precise value of the resistor is not critical.

For example, when using an LED as an indicator light to show that 5 V power is on, we don't need much current from the LED. A current of 1-3 mA is plenty to make the LED visible in room light.

If the supply voltage is 3.3 V and the LED is red with a forward voltage drop of 2 V, the resistor value we need for 1 mA is

$$R = \frac{5V - 3.3v}{1mA} = 1.7k$$

Any value less than about 2k ohms is probably going to be a perfectly fine current limiting resistor. A value of 1k is usually chosen because they are readily available, and it means only 1 value needed in inventory.

Many circuits use resistors in combination with other components to achieve some performance metric. For example, in a resistor voltage divider, two resistors are used. Their relative values are important to reach a voltage divider, not their absolute values.

Or, in an RC circuit, it is the product of the R and C values that creates a specific time constant.

In principle, even a 1 ohm and 10 ohm resistor will give the same voltage divider as a 10k and 100k resistor. Or, if you need a 1 sec time constant, you could use a 1 ohm resistor and a 1 F capacitor.

In practice, most devices limit the amount of current they can easily drive. Unless there is a strong compelling reason and you review the datasheets of the components involved to change your judgment, use resistor values in the range of 1k and 1 Meg. This way in a 5 V system, the typical currents will be in the 5 mA to 5 uA range, values easily sourced, or sunk, or sensed by active devices. When using values outside this range, verify the values are appropriate for the rest of your circuit.

If you need a 1 sec time constant using a 1 k resistor, you would need a 1000 uF capacitor. If such a large value is not readily available, you must find the right compromise between the capacitor values available and the resistor values. A value of 100k ohms and 10 uF might be very reasonable values to achieve a time constant of 1 sec. Of course, this assumed the circuit reading the long time constant voltage has a high impedance compared to 100k.

Power dissipation. The power dissipation of a component is the maximum rated power it can dissipate based on some assumptions about its in-use thermal environment and conducted or convective cooling and the maximum acceptable temperature rise. The rated power dissipation of a resistor is related to the physical size of the component. The larger it is, the more surface area it has and the more heat energy can flow out for a maximum temperature.

An 0603 component is rated at about 1/8 watt. An 0805 is rated at 1/5 watt and a 1206 is rated at about 1/3 watt. If you need more dissipation than 1/3 watt, a larger-size package will be needed. Alternatively, multiple resistors could be used in parallel or series with the required equivalent resistance.

The power consumption in a resistor or other component is the power it generates and turns into heat. The power consumption in a resistor, the power it generates, is related to

$$P = \frac{V^2}{R}$$

This means in a 5 V system, a 50 ohm resistor will consume and generate no more than 0.5 watts. If the duty cycle of the signal is 50%, the average power consumption would be only 0.25 watts. A 1206 resistor could probably handle this level of power dissipation. If you use smaller than 50 ohm resistors, their power dissipation rating will influence the size you should select.

If you need to dissipate more than 0.3 watts, consider using two or three resistors in series or parallel to spread out the power consumption.

When searching for a resistor, we might select:

Package size. This is where using a 1206 or 0402 part is important to know about.

Quantity should be selected as 1 or 10, depending on how many you expect to purchase.

In stock is important for shorter lead time.

> *Watch this video and I will walk you through finding the one resistor in a million to select.*

18.5 Selecting Capacitors

When selecting capacitors, there are a few special criteria to use.

Voltage rating. The voltage rating is related to the reliability and breakdown limits of the capacitor. Generally, the lower the voltage rating, the thinner the dielectric layers inside the capacitor and the higher the capacitance available in the same part size.

In addition, the capacitance varies with applied electric field. The thicker the dielectric layer, the lower the electric field for the same voltage. When the applied voltage is less than about half the maximum rated voltage, the capacitance of the capacitor is close to the rated voltage.

As an important safety criterion, and to achieve close to the full rated capacitance, use a voltage rating that is 2x the application voltage. If you select a voltage rating 2x the application voltage, the part will be more reliable and its capacitance will be closer to the rated capacitance. If your application is a 5 V system, select a capacitor rated for at least 10 V.

Dielectric material, such as X5R or X7R, is usually not critical. This will affect how the capacitance varies with voltage and its temperature sensitivity.

Low mounting inductance. For decoupling applications, having low mounting inductance is important. This means using a MLCC capacitor is valuable. With an MLCC capacitor, it is possible to engineer a mounting inductance as low as 2-5 nH in a 2-layer board.

But the mounting inductance of an MLCC part is more about how it is connected into your board from its pads to the pads of the IC than the MLCC part size itself. **Figure 18.2** shows a few examples of mounting methods for 1206 capacitors. The mounting inductance in these examples ranges from 5 nH to 0.5 nH.

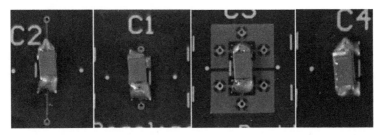

Figure 18.2 Examples of different mounting geometries for decoupling capacitors.

Do not use a leaded capacitor of any sort for decoupling where low mounting inductance is important, as these will have way too much lead inductance. A leaded part, even a small size, will have a mounting inductance on the order of 10-50 nH. This is just because of the length of their leads and their internal construction.

There are four classes of capacitors based on their dielectric material.

- ✓ MLCC, using ceramic dielectrics
- ✓ Tantalum, polarized
- ✓ Electrolytic, polarized
- ✓ Other, nonpolarized dielectrics like mica, paper, polymer, and Teflon

18.5 Selecting Capacitors

Avoid using a *polarized capacitor* unless you need the higher capacitance. When mounted on the board, if you accidently reverse the capacitor, at the least, it will end up looking like a short; at worst, it might *explode*. The end result of connecting a polarized electrolytic capacitor incorrectly, is shown in **Figure 18.3**.

Figure 18.3 The aftermath of an electrolytic capacitors connected the wrong way to the power and ground connections.

For VRM filtering, using a large-value, leaded, polarized capacitor is sometimes acceptable. The higher loop inductance in series with the leaded capacitor may be a small hit. An electrolytic capacitor has the other advantage of a higher ESR than a comparable MLCC capacitor. This means more damping of any parallel resonances.

When we start with 1,000,000 options, we use the criteria we know are important in our application first, and then zoom in on those.

Usually, these criteria should limit the number of parts from which to select from 100,000 to less than 200. In browsing through the list, you can get a feel for the price-performance trade-off and select the lowest-cost part with the values that are acceptable, available in the quantity you want and in the time frame you need.

> *Watch this video and I will walk you through finding the one capacitor in a million to select.*

18.6 The BOM

If the library used by your PCB design tool is well documented, you can generate a final BOM automatically from the schematic.

Alternatively, if you plan to get all of your parts from one vendor, like www.Digikey.com, if all of your parts are in a cart, you can automatically export a complete BOM with one mouse click.

The BOM should include the major components, links to their datasheets, schematic symbols, layout footprints, vendor, and costs. In the early planning phase, the BOM is the checklist to make sure you can find all the noncommodity parts with acceptable delivery time and pricing. It is assumed that all the commodity parts like resistors, capacitors, and LEDs have multiple vendors and are not under allocation.

Here are the best design practices for generating the preliminary BOM:

1. From the functional block diagram and rough schematic, you should know the major components you want to use.

2. Find a vendor for each part and delivery time and cost.

3. To add a part to the library and integrate it into a schematic, each part you use should have a symbol used in the schematic, and a footprint used in the layout.

4. If you find a part you want to use, consider getting the symbol and footprint from SnapEDA or Octopart. These are often linked from Digikey.

5. Find the symbol and footprint that can be added to the library or create it yourself.

6. As a risk reduction to verify you have the correct footprint, purchase a few units of each new part, and place one on a 1:1 print out copy of the top-layer footprint to verify it fits.

7. Alternatively, for a new part, consider building a breakout board to qualify the part and its footprint.

8. Consider making the pads in the footprint slightly longer than the leads to make it easier to solder.

9. Use the fewest number of unique parts. Reuse a part if at all possible to reduce the inventory requirements and reduce the possibility of mixing up similar-looking parts.

10. Read the datasheets for all noncommodity parts. Sometimes they are ambiguous and may require more research. Sometimes they are wrong or misleading. Use your engineering judgement when you read them.

11. You should be prepared to reverse engineer the properties of a component if the data you need is not available, like the rise time of an output. This is another reason it is good to reuse parts. Leverage the investment you make in researching a specific part in future designs.

18.7 Summary of the Best Design Practices

Here are some tips to consider when selecting the best parts for your design. When searching through Digikey consider these tips:

1. Generally, all the parts you will need should be available as surface-mount devices. These will be the lowest-cost, and the smallest-size parts.

2. You should feel comfortable assembling 1206 parts and maybe 0805 parts. If you had a choice, the lowest risk and highest chance of success, unless you have a strong

compelling reason otherwise, use 1206 parts for capacitors and resistors when planning for manual assembly. Be aware of the parts you have in inventory.

3. If you use a component that will connect to something off the board or may have some mechanical stress on it, ALWAYS use a through-hole part. This will provide extra mechanical strength to the part and minimize the chance of pulling the part off the board.

4. For decoupling capacitors, use the largest capacitance that will fit at a reasonable price in the smallest SMD device you are comfortable assembling with a voltage rating 2x the DC voltage of the rail. A typical value is 22 uF 10 V in a 1206 package.

5. Some specialized components such as sensors or display devices are only available in through-hole parts. Consider using socket pins on the board into which the component will be inserted rather than directly solder the component into the board.

6. As a general rule, try to use the fewest number of unique parts in your design. Gain confidence in the specific parts and reuse them in future designs. When you use a part and it works, add it to your favorites list on Digikey and select it as a first choice in your next design when appropriate.

7. The cost of a part is related to its volume use. Sometimes common parts values, like 1k ohm, are cheaper than 1.2k ohm parts, for example. Be prepared to use a value slightly off from what you need if it is significantly cheaper and the specific part value is not critical.

8. When selecting IC packages for manual assembly, do not select a No Lead part, as these will have pads under the package and be very difficult to align to the board pads. It requires a special pick-and-place tool. Use an IC with leads.

9. When planning for manual assembly, when possible, use a package with a lead pitch of 50 mils. This is generally an SOC

or SOIC style package. It will be easier to assemble than a package with a 25 mil lead pitch, like a TSOP.

10. When searching for a specific part, only use filters for specific features you need. Keep the part values in a range.

11. Always search for parts that are in stock, active, and available.

12. Always search for parts that have datasheets available. It is critical to have the datasheet available for every complex part you intend to use. A PDF copy should be placed in a convenient folder or printed out. Many EDA tools have links to the URL of the datasheet that can be accessed directly from the symbol on the schematic diagram.

18.8 Selecting Parts for Automated Assembly

In some board design projects, the fab vendor you select will also do the assembly. You can supply a BOM that the vendor uses to purchase the parts used on your board.

Alternatively, some vendors, especially the lowest-priced Asia vendors, have a special relationship with specific distributors. They will only use the parts from these vendors. This means that when you select parts to use in your design, you must be aware of the vendor's requirements.

If the assembly vendor has restrictions on the parts they will use, select the parts for your design from their options. Be aware of which parts are in stock and available when you send your board for fab.

For example, JLCpcb uses LCSC as their exclusive parts distributor. Some parts are commonly used and considered BASIC parts. There is no premium to use BASIC parts.

Other parts are not as readily available. These are extended parts. To assemble an extended part to your board, someone must go to the back storeroom, grab a reel of these parts, and connect it to the pick-

and-place machine to assemble on your board. This is an extra cost per extended part, and there are a limited number of openings on the pick-and-place machines for additional extended parts.

When selecting specific parts to use on your board, be aware of the impact your part choices have on availability and cost.

18.9 Practice Questions

1. What is a commodity part?
2. What is the typical resistor tolerance to choose?
3. What are two examples of noncommodity parts in a design?
4. What is the first and most important criterion when selecting a part you will order soon?
5. When should you look for availability of noncommodity parts?
6. What information should be in the BOM?
7. What is a common capacitor value and specification for use as a decoupling capacitor?
8. Why should you never use a through-hole capacitor for decoupling?
9. What voltage rating should be selected for a capacitor?
10. How much power can a 1206 resistor dissipate?
11. What is the difference between power consumption and power dissipation?
12. The 50 ohm resistor in the front of an oscilloscope cannot handle more power consumption than 0.5 watts. What is the maximum input voltage to never exceed with a 50 ohm input?

Chapter 19
Step 3: Schematic Capture and Final BOM

The schematic identifies all the components and their connectivity. It says nothing about performance, ONLY connectivity. Once connectivity is established, everything else is about noise, which the schematic says nothing about. This is where the physical design is so important.

> *Signal integrity, cross talk, power integrity, and EMI all types of noise — live in the ideal wires and the white space of the schematic.*

Before you start the formal schematic, hand-draw or sketch the functional block diagram, then draw the schematic roughly by hand to understand the circuit. This makes drawing the real schematic so much easier.

The schematic used as input to a layout tool usually has some different features than a schematic used to describe a circuit to a simulator.

We refer to the schematic used as the input to a layout as the *schematic capture*. This is not necessarily meant to look the same as a circuit description. All schematics will identify the components, their parameter values, and their connectivity. However, in a schematic capture, the format can be different. Net names are often used to provide connectivity across the schematic sheet or sheets. It simplifies the use of a complex array of wires that may be difficult to follow.

The guiding principle in creating a schematic is that it should be easy to read by a fresh pair of eyes. The easier it is to understand the details of the schematic, the less risk of making mistakes.

Generally, this means there should be a structure to the schematic. The same block diagram for the function can be used to organize the schematic into blocks or even pages.

Usually, power flows from an input in the upper left corner through the rest of the design. When possible, the signal flow should follow some pattern, like left to right or top to bottom.

Use text wherever possible to label blocks or special functions.

The design process is a creative process. But it is not about synthesizing something from nothing. The most efficient design process takes advantage of existing building blocks. Think of the system you want to end up with as built from smaller functional units.

While there are literally thousands of building block elements to choose from, there are a handful of commonly used, really valuable circuit elements important to have in your toolbox and get comfortable using.

The more building block circuit elements and components you know about, the more pieces you have available from which to build your system.

The schematic is complete when all the components are added and routed, an ERC has been performed and passed and a design review is completed. At this point, the design is moved to the layout phase, Step 4.

19.1 Picking a Project Name

The very start to the schematic capture is to pick a name for your project. While it is such a simple process, there are some bad ways of naming projects, and some good ways.

Generally, in most EDA tools, the overall board project has a name and each schematic sheet and the layout file have names. Your project name can be your board name.

Avoid a name for your project like: board1 or testboard. You will forget what this design was next week. You will get it confused with the next three boards you design. This is a bad habit that will absolutely increase the confusion (risk) of mixing up boards. You should pick a project name that will be unique and scalable to other boards, either as revs of this board, or related.

I recommend using a file name for your project that is self-documenting. It can include for example:

- ✓ Your initials so you can tell at a glance this is a project you created.
- ✓ A brief description of the board name.
- ✓ A date code of when it was created.

Here are some examples:

Elb_practice1_2021-01

Elb_SMAlaunch_2020-12-21

Elb_SAMD21_2020-09-01

This way, you may have a few related board projects, but they could have different date codes.

Within the project, there will be schematic sheets and layout sheets. These files will remain within your project folder so they can be specifically about this project.

When you make a change, you can either save with the same name and overwrite, or add a rev1, rev2, etc. to the sheet name. Sheet names can be used over and over again between different projects, as they will always live within the project folder. They could be 1_*power* or 3_*ucontroller* or 7_*ADC interface*, for example.

Adding a number to the front of the sheet name will make it easy to keep track of each sheet when adding port names. A port is a connection between nodes on two different sheets. If each sheet has a unique number associated with it, a port name would just have to refer to the sheet number and net name to connect to it.

While it is rare, it sometimes happens that a file gets corrupted. If you put 10 hours into creating the file and it gets corrupted, you don't want to lose all of your work. It is valuable to periodically save a file under a new rev or date code so you have a recent version to go back to. Your prjPCB database will keep track of the latest saved filename when you open your project again.

If you save different revs of your schematic sheets, your project file will keep track of the latest one. After you have saved the new sheet name, you can delete the older revs at your leisure using a file browser tool.

19.2 Schematic Capture

The schematic identifies all the parts that will be used in the design and their connectivity in the circuit.

Before you can use a part in a schematic, it must be in the library. If a part is not in your library, you cannot add it to the schematic.

If you have a global library from your librarian, you can get most of your parts from this library. You may have to add a part not in your global library to a local library tied to your specific board project. This is a hybrid approach, using a global library and a local library.

Plan on periodically sweeping all new parts into the global library after the symbols and footprints have been verified.

The schematic is the electrical design. At the end of this step, all the parts should be identified and added to the schematic page as symbols.

All the terminals should be connected to all the correct nets and the nets labeled. The BOM should be complete with the list of ALL the specific parts, using the fewest number of unique parts.

19.3 Take Ownership of Reference Designs

There are thousands of designs posted online for projects ranging from simple LED resistor circuits to high-performance computer motherboards. They are always useful to take a look at to gain insight and see how others approached a problem you may be looking at.

However, do not blindly accept recommendations from reference designs. Just because a reference design is published and it may "work" does not mean it uses best design practices and should be replicated. Oftentimes, an eval board is designed by a summer intern without much more experience than you.

Once you decide to use a reference design, it becomes your design. Take ownership and take responsibility for the design.

An important issue is decoupling capacitor selection. They are often arbitrary and often wrong in many reference designs. A reference design may work in spite of the design decisions, not because of them.

If you are using a reference design to base your design:

- ✓ Challenge all the assumptions.
- ✓ Don't assume the person who published the reference design knew what they were doing.
- ✓ Make it your own, take responsibility for it.
- ✓ Check the design against datasheets.
- ✓ Do not carry through legacy assumptions in the design that do not apply anymore.
- ✓ Keep in mind that oftentimes, designs work in spite of the decisions made, not because of them.

19.4 Add Options to Your Schematic

The design should be verified and checked for consistency with the POR, correct functioning, and correct connectivity.

The power delivery design should be evaluated for current handling, power consumption, and decoupling capacitor usage.

If you are not sure if an enable pin, for example, should be connected to a HIGH- or LOW-level voltage for a part and for your design, do not guess. Read the datasheet carefully, ask someone else that has used the part, google it, or buy a few units and test it out on an eval board.

> *Remember, just because you find an answer online does not mean the answer is correct!*

If you still do not have an answer you are confident with, flag this as a risk site and spend additional time investigating the answer. If you release your schematic not sure of the answer, Murphy's law says you chose the wrong answer. This is why it is so important to reuse parts you have already used. You have confidence in how they work.

As a final alternative, if you do not have confidence in your answer at the time you have to go to board fab for your prototype, add a three-pin header switch that would connect the enable pin through a resistor to either a HIGH or a LOW voltage. Use this prototype to definitively answer the question for the next time you design with this part. This is part of the risk mitigation process.

19.5 Best design practices for Schematic Entry

An expert is someone who has made all the mistakes possible. By following these best practices, you will avoid some of the common mistakes that others have made so you do not have to make them for yourself.

There is a difference between what to implement and how to implement it. The features to implement are tool agnostic. Some

19.5 Best design practices for Schematic Entry

features are easier to implement with some tools. Different tools will have different methods to implement a feature.

Here are the best design practices for generating the schematic capture:

1. "The difference between a good schematic and a bad schematic is in how easy it is to understand," Mike Hurowitz, an expert PCB designer.

2. Don't make the reader work at understanding the circuit diagram — make it easy to read.

3. Keep it neat.

4. An orderly schematic translates into an orderly layout. Use the schematic as a guide for the layout.

5. Label the board with a descriptive name and use revision control on the schematic page.

6. Use one schematic page if it fits and is neat.

7. If the entire schematic does not fit on one sheet, partition it into functional blocks and place each functional block on a different sheet.

8. If you use multiple sheets, add a sheet name to the schematic sheet at the very top, number your sheets in their names, and add the sheet number in front of each port to which it connects.

9. The schematic should be laid out in an orderly pattern, with some consistent flow, usually from upper left, down, or across.

10. If you plan to probe a net, or its routing is critical, be sure to use a special net name. Use logical words, not just numbers.

11. Net name conventions:

    a. A power net should start with a + or a -.

b. Never use a space in a net name. Use camelCase or a dash "-", or an underscore "_".

c. Don't use decimal points — they are too hard to see.

d. Use a v to separate numbers as +3v3.

e. All power rails in a series path should start with the same values and then an "_" and some description like +5v_in or +5v_afterFilter.

f. Don't label a net as just TX or RX. It will be a TX on one end and an RX on the other.

g. When a net comes from an IC, use its name in the net label, like FTDI_TX or FTDI_DTR.

h. Make net names in all caps. Makes it easy to identify a net name.

i. When adding the same net name to a different trace to provide connectivity, either copy and paste or select the net name from the pull-down menu to minimize the chance of typing in the wrong net name.

j. All wires with the same net name are connected together in the database.

k. Make wire lengths at least as long as the net label.

12. If a net goes off the schematic page to another page, use a port on each page to make the connection. The port names on both sheets should be the same and the same as the net names on each sheet.

13. For global power, use a power port to connect to the terminals.

14. The same connected traces all belong to the same net. If you need to have the same connection with different names, like analogVCC and coreVCC and I/OVcc, consider using a 0 ohm jumper in series, but not between the IC pad and its

decoupling capacitor. This splits the net into two different nets. If an analog Vcc, consider adding a ferrite bead in series with the input voltage.

15. Use a grid of 100 mils for placing parts or routing traces in the schematic. Most symbols have their terminals on 100 mil grid.

16. Add at least a 100 mil long wire segment to each terminal before you connect a component to it.

17. If you cannot make connections between terminals, it is probably because the symbol terminal grid is different from the routing grid. Try reducing the grid pitch to 5 mil or 1 mil until you can make a connection, then immediately switch the grid back to 100 mil.

18. Avoid crossing wires on the schematic. It makes it difficult to tell at a glance what is a crossover and what is a connection. Use net labels to create the connectivity instead of crossing wires.

19. If you need to connect four wires together, do not connect them all at the same cross point. This looks too much like wires crossing over each other. Space the connections apart so you can clearly see each of the connections.

20. By convention, ground symbols point down on a schematic page.

21. Place the decoupling capacitors for each IC in proximity to the IC on the schematic.

22. Where possible, use a symbol for the component that shows some function so it is easy to identify each pin and how it should be connected.

23. Constantly check for connection errors.

24. Place parts associated with each other in close proximity to each other, like test points, indictor LEDs, and switches, ...

25. Do not put all the decoupling capacitors on one page, or all the indictor LEDs or all the test points all on one page. Distribute them close to where they would be in the layout.

26. Try to separate your circuit into logical blocks labeled with text on the schematic page.

27. Avoid the common mistake that a TX from one part connects to the TX of another part. Usually a TX pin connects to an RX pin.

28. If you want to have the option of running a device at 3.3 V or 5 V to experiment, consider adding a voltage selection jumper using a 3-position header and 2-position shorting flag.

29. If you use a reference design from somewhere else, take ownership of it. Understand every detail and don't assume the designer is any smarter or more skilled at circuit design than you are. Mistakes get into reference designs ALL THE TIME. Reduce the errors and bad practices so they don't propagate in your designs.

30. Feel free to modify a reference design using your own judgement. Never use the excuse, "It was in the original design." For example, most designs recommend three different value capacitors, or values that are 0.1 uF. This is totally misleading.

31. Keep the BOM count of unique parts to a minimum. Use a 1k resistor in place of a 1.2 k resistor when you can. Balance final performance with reduced number of parts. This will reduce the assembly cost, reduce the inventory costs, and reduce the chance of making a mistake in the assembly.

32. Do not arbitrarily change a value of a component in the symbol after placing on the schematic page. Be aware of how this component will be displayed in the BOM.

19.6 Design Review and ERC

After you think the schematic is done, take a look at it. Here is the checklist you should use when reviewing your design and someone else's design. Every time you encounter an error you've made, add it to this list so that you do not make the same mistake next time.

- ✓ Does it include all the functional parts it should?
- ✓ Are all the parts connected up correctly?
- ✓ When there are two specific connections like D+ and D- pins, or TX and RX pins or SDA and SCL pins, are you sure you connected them up correctly and not opposite (like Captain Murphy's technician)?
- ✓ Is it easy to read?
- ✓ Is it laid out in a neat, orderly fashion?
- ✓ Can you tell the functioning of the circuit from the schematic?
- ✓ Are nets you will want to probe during debug labeled well?
- ✓ Do you have test points, indicator lights, and switches, which will aid in debug, in proximity to what they connect to?
- ✓ Are the decoupling capacitors for each IC in close proximity to the IC?
- ✓ Is every component labeled with a reference designator and displayed on the schematic?
- ✓ Does very component have an associated footprint?
- ✓ Have you done everything to minimize the number of unique parts?
- ✓ Have you used a minimum number of libraries from which to pull parts and are these libraries saved in a file in proximity to your board design project?

19.7 Practice Questions

1. What are two guiding principles when drawing a schematic sheet?

2. What is an example of a good project name?

3. What is an example of a bad project name?

4. What is the value of a reference design?

5. What is the danger of using a reference design?

6. What is the most important action you should take when using a reference design?

7. If you are not sure which of two options to use in your schematic, what can you do?

8. What are three common mistakes you could probably make?

9. What information will be in the BOM?

10. What are three examples of best practices for schematic entry?

Chapter 20
Step 4: Layout — Setting Up the Board

After the schematic has been completed and the ERC passed, it's time to start the layout. This is where the schematic is turned into a physical design, where signal integrity is important, and the board is sent out for manufacturing.

It is in the layout phase where the physical implementation of the circuit is created and where signal integrity, power integrity, and EMI can play an important role. This is when the ideal wires and the white space of the schematic are turned into physical interconnects.

> *Signal Integrity, cross talk, power integrity, and EMI, all types of noise — come alive in the layout.*

The output of the layout are the design files that are sent to the manufacturer to fabricate the board and assemble the components.

20.1 Layout

In the layout phase, the following steps must be completed:

- ✓ Create a PCB file
- ✓ Create the board shape, the outline, and the keepout layer
- ✓ Establish the design rule constraints
- ✓ Create the stack-up
- ✓ Place the components
- ✓ Route the components
- ✓ Complete a DRC

- ✓ Prepare the documentation
- ✓ Send the files out for manufacturing

All parts are placed, routed, and labeled for:

- ✓ Design for connectivity (DFC)
- ✓ Design for manufacture (DFM)
- ✓ Design for reliability (DFR)
- ✓ Design for assembly (DFA)
- ✓ Design for bring-up and debug (DFB)
- ✓ Design for test (DFT)
- ✓ Design for user experience (DFU)
- ✓ Design for performance (DFP): DC resistance and current handling
- ✓ Design for performance (DFP): Signal and return path routing
- ✓ Design for performance (DFP): Power distribution routing
- ✓ Design for performance (DFP): EMC

The Gerber files and NC drill file are exported and sent to the fab vendor for production.

Any assembly information and BOM are also sent to the assembler.

20.2 Board Dimensions

Be aware of the terms we use to describe specific features of the traces on the board. We sometimes refer to a copper conductor on the board used to route between two or more terminals as a *trace* or a *track* or as *etch*. The latter is an older term.

Line width refers to the cross-sectional width of the signal trace in the plane of the board. This is at 90 degrees to the long axis of the trace.

Trace thickness refers to how thick the conductor is in the vertical direction. It is measured in mils or oz of copper per square foot.

The *spacing between signal lines* is the edge-to-edge distance apart.

The *pitch between signal lines* is the distance between the center to center of two adjacent traces.

The *dielectric thickness* is the thickness of the insulating layer between the bottom of metal layer 1 and the top of metal layer 2.

The *board thickness* refers to the overall thickness of the board from the outer surface of the metal on layer 1 to the outer surface of metal layer 2. It does not include a soldermask or silk screen layer in the thickness specification.

The board thickness usually comes in specified values. This is so that there is compatibility with the fabrication tooling during manufacturing and end-use applications for connectors or physical hardware with which the edge of the board needs to be compatible.

The board thickness you use also affects the stiffness and mechanical strength of the board.

The standard thickness is 62 mils (1.6 mm). Boards also come in thicknesses of 93 mils (2.4 mm) and 31 mils (0.8 mm). Much thinner than this and the board may be more flexible than you expect and may warp.

When there is a solid plane of copper on the bottom layer and a few traces on the top layer, a board thinner than 1.6 mm may warp or twist like a potato chip due to the asymmetry in metal and the different thermal expansion coefficient of copper (18 ppm/degC) and the epoxy glass (35 ppm/degC).

When it is thicker than 62 mils, it may not fit into the final product you want, or header pins may not fit all the way through the board as you expect.

Unless you have a strong compelling reason, a 62 mil thick board is a good option.

20.3 The Layers in a Board Stack

In Altium Designer, for example, under the menu item Design/Layer Stack Manager, the stack-up of the board identifying each of the layers can be found. An example for a 2-layer board is shown in **Figure 20.1**.

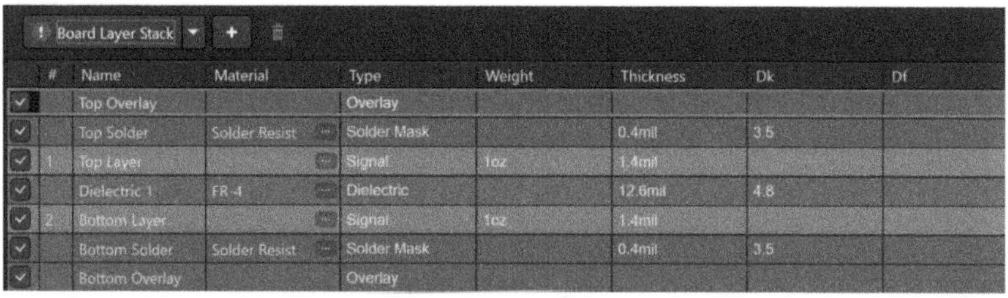

#	Name	Material	Type	Weight	Thickness	Dk	Df
	Top Overlay		Overlay				
	Top Solder	Solder Resist	Solder Mask		0.4mil	3.5	
1	Top Layer		Signal	1oz	1.4mil		
	Dielectric 1	FR-4	Dielectric		12.6mil	4.8	
2	Bottom Layer		Signal	1oz	1.4mil		
	Bottom Solder	Solder Resist	Solder Mask		0.4mil	3.5	
	Bottom Overlay		Overlay				

Figure 20.1 Example of the layer stack for a 2-layer board.

If you can fit your routing on two layers, with the bottom layer as the ground net, use two layers. It will be a lower-cost board. In addition to the copper layers, every board will also have a soldermask layer on the top and bottom and a silk screen layer, often called the overlay layer.

If you need additional routing layers, use four layers. Due to the manufacturing process, board layer count only comes as even numbers. Never select an odd number of layers in a board. As described in a later chapter, the recommended stack-up for a 4-layer board is a signal-power/gnd/gnd/signal-power layer.

20.4 Negative and Positive Layers

When the features used on a layer appear in the layout screen in the EDA tool, as they would appear in the as-fabricated board, we call these layers positive layers.

20.4 Negative and Positive Layers

When we route a trace on a board, we see the actual path the copper trace will look like as the colored line. What we see as traces on the screen is what we get as traces of copper in the final board. This is a positive layer.

When we see a soldermask layer where there is color, there will be a hole or empty space in the soldermask on the board. The soldermask layer is a negative layer.

The same is the case for the stencil or solder paste layer. Where we see colored features in the solder paste layer, there will be openings in the stencil and solder paste will print through onto your board. The solder paste layer is a negative layer.

It is sometimes tempting to select the ground layer in a board stack-up as a "plane" layer. After all, we intend to use this layer as a ground plane.

However, the definition of the "plane" setting in most EDA tools is that it is a negative layer. While we don't have to do anything special when using this plane except connect it to the ground net, it will look really weird when displayed in the EDA tool.

Only open regions in the plane will appear as colored regions in the artwork for that layer. This means solid-colored regions of the ground layer are really open regions. It is very difficult to apply our visual intuition and debug the board and check for gaps if the ground layer is a negative layer.

Unless you have a very strong compelling reason otherwise, always select the ground planes as signal layers so they are routed as positive layers. You will have to manually add a polygon pour of copper fill on this layer and connect it to the ground net, but you only have to do it once. The benefit of being able to see the ground plane as it will be fabricated will reduce the risk of making inadvertent gaps in the return path, which is worth the added time.

It is possible to set up the preferences in most EDA tools to automatically adjust the copper pour to the ground layer if you add a cross-under trace in the copper layer. Just be aware of the condition to minimize the length of any cross-unders in the copper pour of the bottom ground layer to reduce the ground bounce switching noise.

To reduce the chance of adding ground bounce to your board, you always want to reduce the number of cross-unders, and when they are used, keep them short.

20.5 Examples of Some Fab Shop DFM Features

In the PCB industry some fab houses can do as narrow as 2 mils, but they charge more. Some can do 4 mils as standard, and some 5 mils wide as standard, with no cost premium.

Many of the low-cost fab vendors have a minimum line width, for no additional charge, of 6 mils. Check out the website for PCBway.

As shown in **Figure 20.2**, for PCBway, the narrowest line and space at no additional charge is a 6 mil wide trace and 6 mil space.

Figure 20.2 Example of the features PCBway is capable of fabricating at no extra charge.

On their website, the thickness refers to the total thickness of the board. The default value of 1.6 mm is about 63 mils thick. This is a standard thickness. The minimum track/space is the line width and space between metal features. They can do less than 6 mil/6 mil, but it will cost more. You can play with the calculator to see how much more it will cost.

> *Unless you have a strong compelling reason otherwise, do not use a line width or other metal feature smaller than 6 mils or a space between two metal regions less than 6 mils or you will be charged extra. Remember this!*

Likewise, be sure to keep the maximum board size less than 100 mm x 100 mm, which is 3.94-inch x 3.94 inches. If your board is larger than this, the price will jump from about $1 per board to more than $10 per board. Just to add a little margin, as a best design practice, always use a board outline less than 3.9 inches x 3.9 inches. Even safer is 3.8 in x 3.8 in.

The minimum hole size refers to the via drill size. The narrowest hole for no additional cost is 0.3 mm or 11.8 mils. I recommend, as a best design practice, a via drilled hole size of 13 mils unless you have a strong compelling reason otherwise. This provides a little margin and provides a safety margin if you use a fab vendor with a slightly larger, lower limit.

Is there a downside to using a 6-mil wide trace all the time? What might you imagine as a potential problem?

The series resistance of a 6 mil wide, 1 oz copper trace is less than 0.1 ohm/inch and the maximum current capacity with no danger of heating is about 1 A. Unless your application requires more than 1 A in a signal line, always use the narrowest line as this will make routing easiest.

Likewise, use the smallest via size, a 13 mil drill hole, to enable easiest routing.

Be aware of the limitations of the fab shop you will use. Before you begin your design, visit the fab vendor's website to identify:

- The minimum line width and space
- The minimum via drill hole side
- The maximum board size at the lowest cost

Many fab shops have price break points. If you want a feature smaller than these limits, or a larger board or more layers, they will fabricate it, but will charge you more for it.

If you can implement the capabilities you need for your project with the lowest-cost board, why would you want to pay extra if there is no value returned?

Here are examples of four board shops with different lowest-cost process features. Take a look at their websites and evaluate their lowest-cost features.

https://www.pcbway.com/

https://jlcpcb.com/

https://www.4pcb.com/pcb-prototype-2-4-layer-boards-specials.html

https://docs.oshpark.com/services/

As a general rule, all fab shops can do the following at no extra charge:

- ✓ Minimum feature size: 6 mil wide line 6 mil space
- ✓ Minimum drill size: 13 mil drill diameter
- ✓ Maximum board size < 100 mm x 100 mm or 3.9 in x 3.9 in

We will stick to these feature sizes unless there is a strong compelling reason otherwise.

20.6 Setting Up Design Constraints

Use the constraint manager in the design rules pull-down to set the minimum trace width to 6 mils and the maximum to 20 mils.

The minimum trace separation should be 6 mils and the via drill diameter is 13 mils.

As a safe margin, a spacing of 10 mils between metal features can be set in the constraints. If you can get away with this large a gap, it may improve your board's reliability. The IPC 2221b specification recommended maximum voltage between adjacent conductors for a 6 mils spacing is 50 V.

If you need the tighter space for routing, you can reduce the constraint to 6 mils.

The capture pad for a drilled via should have a minimum annulus of 6 mils. This is the narrowest metal feature. If a 6-mil annulus is around a 13 mil drill hole, the outer dimension of the capture pad is 13 mils + 6 mil + 6 mil = 25 mil diameter. This is sometimes referred to as the via diameter, when in fact it is the diameter of the capture pad around the via.

20.7 Thermal Reliefs in Pads and Vias

There is an important consideration when designing pads and vias related to thermal management. Any component that will be soldered to the board will sit on a pad. The pad will be isolated from the rest of the metal of the net with soldermask. This soldermask prevents the solder from spreading out on the copper net, depleting the pad of enough solder to make a reliable solder joint.

However, the rest of the copper connected to the net will also affect the thermal transfer from the pad. This is especially important when hand soldering a component to a pad.

If there is a lot of metal connected to a pad, the heat from the tip can be sucked out by the thermal sink of the rest of the copper and it can be very difficult to solder a component to the pad. This is especially the case when the pad is connected to a large copper pour polygon.

While the soldermask prevents the solder from wicking onto the rest of the copper net, we need to prevent the heat from wicking onto the rest of the copper on the net.

We do this by shaping the copper path from the pad to the rest of the net. If we make the connection path very narrow, we can keep the electrical resistance low, but make the thermal resistance high.

These specially designed regions that thermally isolate a component pad from the rest of the copper in the net are called thermal relief pads. They are an isolation moat around the copper pad, with a few drawbridges of copper to make it easy for current to flow, but hard for heat to flow. **Figure 20.3** is an example of a thermal relief pad.

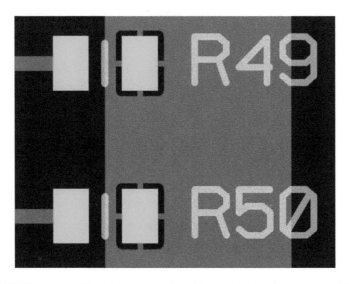

Figure 20.3 The pads in the column are in the middle of a plane, but separated from the plane using thermal reliefs.

At first glance, it looks like the resistance should be very high. After all, the current has to neck down through very narrow traces. The way to evaluate the resistance through these narrow drawbridges is by putting in the numbers.

The resistance of a trace is the sheet resistance x the number of squares. The sheet resistance of 1 oz copper is 0.5 mohms/square. A glance at the narrow sections in the thermal relief pads shows 1 square of metal in each drawbridge tab. This means the resistance of each narrow trace is about 0.5 mohms/sq x 1 squares = 0.5 mohm.

Each small tab has a resistance of about 0.5 mohm. There are four of these short traces in parallel. This means the net resistance from the inside pad to the rest of the net is ¼ x 0.5 mohms, or 0.125 mohms. This is a very small amount and is insignificant.

Even though the electrical resistance between the inside and the outside of the pad is very small, the thermal resistance is high enough to enable robust soldering by soldering tip or by reflow oven.

The width of the narrow traces should never be narrower than the minimum feature size at no additional charge, typically 6 mils. In the example above, the width of the narrow traces is 6 mils and their length is 6 mils.

This is such a common feature to add to a pad that all EDA tools have an automatic feature to turn any pad into a thermal relief pad. Be sure the constraint manager is set for the minimum line width 6 mils or wider.

Vias can also be engineered with a thermal relief to reduce the heat flow from a via to a plane to which it connects. An example of a via with thermal reliefs is shown in **Figure 20.4**.

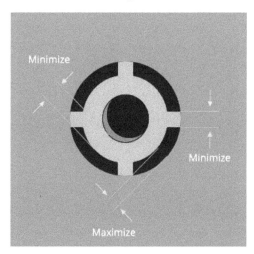

Figure 20.4 Example of a thermal relief via.

Heat flow from the via into the plane is only an issue when you need to solder a pin in the via. If the via connects to a plane and you do not use a thermal relief, soldering the pin in the via is very difficult. The heat will be sucked from the via into the plane, reducing the temperature of the pin and preventing the solder from melting and reflowing.

This is when a thermal relief via is essential. The thermal relief via will be electrically connected to the plane, but thermally isolated. The pin will easily heat up from the soldering iron and it can be easily soldered to the via.

When a via is between a signal trace and a plane and there are no solder pads adjacent to the via, there is no need for a thermal relief via. If the via is connected to a pad through a 6 mil wide trace 20 mils long, there is no need for a thermal relief via. The trace acts as a thermal relief. If a via is between two signal traces on different layers, there is no need for a via with thermal relief.

It is only when the via is inside a solder pad used for the ground connection to a decoupling capacitor, for example, is a thermal relief via needed.

Do not confuse a thermal via with a thermal relief via. Sometimes a via can be used to intentionally suck heat from a component or pad that is connected to a heat-generating component. An array of vias from the pad to a plane can suck the heat out of the pad into the plane. A thermal via does the opposite of a thermal relief via.

20.8 Set Up Board Size and Keepout Layer

If there is no compelling reason otherwise, a convenient board size is 3.5 inches x 2 inches. This is the size of a business card. If done correctly, one of your boards could literally be used as a business card.

As mentioned earlier, if at all possible, keep the board size less than 3.9 inches x 3.9 inches. A larger-size board in any dimension will be more expensive. If you use a larger board size, you should have a compelling reason to pay extra for it. Why pay extra for value you do not need?

Once the board outline is defined, there should be a 20 mil keepout region along the perimeter. This is usually placed on the mechanical layer labeled *Keepout* layer.

The fab shop will use this keepout region to cut out your board from the panel. They will use an end mill drill bit to cut the board. Keep all metal in your design inside this keepout region so that it is not accidently cut by the mill bit.

The size of your finished board will be defined by the middle of the keepout region. This means the closest metal on your board will still be about 10 mils (0.25 mm) from the edge of the board.

If you do not include a keepout layer in your list of Gerber files sent to the fab vendor, chances are they will not fab your board.

20.9 Practice Questions

1. What is a convenient board size to match a business card?

2. What is the typical maximum size board that a fab shop will not charge extra for?

3. What is the typical design constraint for the minimum line width and spacing?

4. What is a typical via drill diameter?

5. When should a thermal via be used?

6. When should a thermal relief via be used?

7. What layer would contain the board outline?

8. What is the width of the keepout region of the board outline?

9. What is the common thickness of a circuit board?

10. Why is it preferred to use a ground layer as a signal layer rather than as a plane layer?

Chapter 21
Floor Planning and Routing Priority

Floor planning is the process of placing the parts on the circuit board. There is a process and recommended order for placing the parts on the board to reduce reliability, assembly, and performance problems.

Once placed, you can begin routing the signal and power connections between terminals of the components. If only connectivity is important, it does not matter how the wires are routed. But as a good habit, if it is free, there are a few simple best design practices that should be followed to reduce the noise in your board.

Even if the interconnects are transparent in your current board and it does not require lower noise, following these best design practices is still a good habit. Use every design opportunity to practice the routing principles you may need for your next design.

21.1   Part Placement

When the parts are transferred from the schematic capture page, they will be placed in a region of the layout page off the board outline, but showing the connections between each of the parts. They are placed in a section of the layout page sometimes referred to as a *room*.

The connections between all the parts that comes over from the schematic are called the *rat's nest or ghost routing*. It sort of looks like the random organization of the twigs and straw and long stems of grass a rat uses to line its nest. A comparison of a real rat's nest, a jumper wire circuit in a solderless breadboard, and the starting place for layout is shown in **Figure 21.1**.

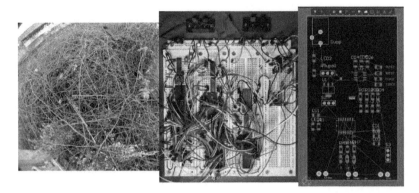

Figure 21.1 The term rat's nest refers to wiring that is connected correctly but has no order or consistency. Note the actual rat's nest on the left, the jumper wire routing in the middle, and the starting place for layout on the right.

When considering the placement of parts on the board, here are some basic guidelines:

1. Connections to the outside world or mechanical issues should be near the board edge.

2. Antennas should be near the board edge. The antenna element should either stick over the edge of the board, or any planes underneath the antenna element should be removed, unless specifically required for the antenna, such as for 77 GHz radar.

3. Place the connectors first, then the wireless parts, then those with special mechanical considerations, then the most sensitive parts, and then the rest.

4. Parts with a lot of connections between them should be placed in proximity to each other.

5. When possible, place the parts so the board looks like the schematic. This will simplify debug when you can easily compare the schematic and the board. For example, if the LED is in series with the resistor, place them in a straight line on the board.

6. Consider rotating parts, or changing their orientation to facilitate the routing and reduce the number of cross-unders needed, as seen in the rat's nest routing.

7. If a test point is designed for a clip-on mini-grabber, it should be near the edge of the board to fit the clip.

8. Test points designed to have a 10x probe needle tip and spring ground tip inserted into them should be 40 mil drilled diameter holes with 6 mil annulus and spaced 300 mil center to center near where they will be measured.

9. Components associated with other specific components should be placed in proximity to each other:

    ✓ Decoupling capacitors should be placed close to the IC they decouple, connected to the IC pins by short, wide paths.

    ✓ LED indicators should be close to what they indicate.

    ✓ Switches should be placed near what they are switching.

    ✓ Isolation jumpers should be placed between the elements they isolate.

    ✓ Small via hole test points should be placed near the lead they are testing, with an adjacent return via connection.

These are a few best design practices for placement of parts. If the floor planning decisions will affect any aspect of the board, we want to follow the best design practice to reduce the risk and minimize the noise.

The floor planning decisions will affect:

  ✓ Mechanical interfaces
  ✓ Thermal management
  ✓ Ease of assembly
  ✓ rf performance
  ✓ Noise coupling

- ✓ Ease of testing, bring-up, and debug
- ✓ User interface

Think about these elements and what you might do about the part placement to facilitate these conditions.

21.2 The Order of Placement and Routing

While there are multiple right ways of doing the routing once parts are placed, there is a methodology that results in the lowest risk, lowest noise, and best chance of success.

All approaches must result in the correct connectivity. This means all the ghost wires must be connected between the appropriate pads.

If there is no ghost wire present between two pads you know should be connected together, this means the schematic is wrong. Go back and fix the schematic.

If connectivity is the only property that matters, it doesn't matter where the parts are placed. But, as a good habit, once connectivity is established, placement and routing can be done to reduce potential noise.

To control noise, the order of placement should be:

1. Connectors, test points, and components that need to be on the board edge go first. Make sure connectors are oriented so that the connections can be inserted from the outside of the board into the the board.

2. Place ESD protection close to the connector on the board to provide maximum protection in both directions.

3. Power generation VRM components placed close to where power enters the board.

4. Current sense resistors placed next in close proximity to where power feeds the board.

5. Test points and indicators associated with the connectors or VRM go next.

6. Highly interconnected parts are placed near each other to minimize the cross-overs.

7. Space IC components with many connections far enough apart to fit the routing between them.

8. Decoupling capacitors are placed as close as practical to the IC pins they decouple.

9. Ground vias are placed on the other side of the decoupling capacitors and with traces as short as practical to the ground via

10. Crystals or other critical signals, like very high-speed or low-level analog signals, should be close to their IC pins.

11. Try to place sensitive analog parts away from noisy components, like power supplies and digital ICs, with high currents.

12. Signal traces are routed next.

13. Power traces are routed after the signals.

14. Finally, consider the assembly operations. If parts are too close together for easy assembly or repair, move them apart.

21.3 First Priority: Ground Plane on the Bottom Layer

The most important design principle to reduce the biggest source of noise is to use the bottom layer as a continuous return path, connected to the ground net.

There are two methods of creating a ground layer in most EDA tools.

You can select the bottom layer in the board stack-up manager and set it as a plane and then as a ground plane. This makes the plane appear as a negative image. Only the holes in the plane appear as colored structures on the layer. This sometimes makes it tricky to interpret features at a glance.

I do not recommend this method. It has too big a risk of letting errors slip through and not be caught by visual inspection.

The second approach is to label the bottom metal layer as a signal layer in the stack-up manager and draw a polygon on this layer assigned to the gnd net. This way clearance holes and gaps will automatically be placed on this layer when a trace is routed on the bottom layer and not tied to the ground net.

The ground layer will appear as a positive in the EDA tool. Holes in the ground plane will appear as holes in the colored layer. *This approach is preferred.* It allows you to see the gaps in the return plane and is easier to read by the fab house.

The best design practices for routing the ground plane are:

1. Make the plane a rectangle covering the bottom layer and tie it to the gnd net.

2. No need to use thermal relief vias to the gnd plane unless the vias are very close to a component that will be soldered to the board or there are holes for pins to be inserted and soldered.

3. First priority, do not route signal traces in the gnd plane. Keep the gnd plane continuous and solid.

4. Second priority, if you have to route a trace in the bottom layer, as a cross-under, keep the cross-under trace length short so that the gap created in the return plane is short.

5. Third priority, it is better for a trace that has to go from one edge of the board to the opposite edge to make multiple, short cross-unders where needed in the ground layer, than one long cross-under path in the ground layer. This keeps all gaps in the ground plane short.

6. Fourth priority, if the gap in the return plane has to be longer than ½ inch, add a return strap cross-over on the top layer

from one side of the gap in the ground plane to the other side of the gap crossing the gap in the middle of the gap. Use a short length strap in the top layer to provide a path for the return current to step over the gap.

7. All gnd paths on layer 1 from a pad to the gnd via should be routed with 20 mil wide traces and kept as short as practical. The minimum length of a trace from a pad to the ground via should be 20 mils so that a soldermask bridge will cover the trace. This will prevent solder from wicking to the via and robbing solder from the component pad. The short trace acts as a thermal relief.

8. Sometimes routing the gnd via to an IC pin can be done with the shortest path by routing inside the IC footprint, rather than to the outside perimeter of the IC.

9. Do not share return vias to the gnd plane. Each component needs its own separate gnd via. When you share ground vias, you increase the overlap of the return currents and increase ground bounce noise.

10. Be sure to use the "repour" feature for this polygon to adjust for all the routing changes after you have added traces to the ground plane.

21.4 Second Priority: Decoupling Capacitors

After the components are placed, the next priority is to control the self-aggression switching noise on the power rail. This is done by reducing the loop inductance from the IC pads to the decoupling capacitors.

1. Make sure all the decoupling capacitors are placed in close proximity to the IC and its VCC pins they are decoupling.

2. Route all the connections between the decoupling capacitors and the IC power pins with as short and wide a power path as practical.

3. Route the gnd connections to the capacitors with as short a trace directly to the gnd via as possible.

4. Never share ground vias with adjacent components (it bears repeating!).

5. Use multiple, low-inductance decoupling capacitors in parallel to provide more capacitance for low-frequency charge storage and to provide even lower loop inductance with the parallel combination of inductances of each capacitor.

6. Consider using power puddles from the IC pads to the decoupling capacitors and ground puddles on the other end if there are multiple decoupling capacitors.

7. Figure 21.2 shows examples of two different layouts for the same three decoupling capacitors. In each case, there are three 22 uF capacitors used. One design on the left has a lot of loop self-inductance, and the other on the right has reduced loop-self-inductance.

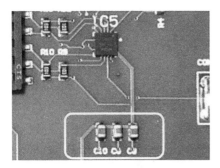 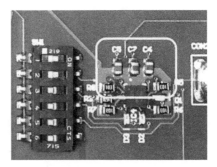

Figure 21.2 Examples of the placement of the same three decoupling capacitors with good and bad layouts. On the left is a larger loop inductance than on the right.

21.5 Third Priority: Ground Connections

The biggest source of signal-to-signal cross-talk noise to reduce is from the loop mutual inductance between signal-return path loops. The first step in doing this is to make all ground connections from a ground pad of an IC to the bottom ground plane with as short a trace as practical.

Sometimes this means routing from the IC pad inside and under the package to the ground via.

Do not share ground vias. There is never a compelling reason to share ground vias. While the impact on the cross talk from a shared ground via may be small, due to the short, overlapping path of the two return currents, it is an easy change to make. If it is free and may decrease noise, it is a good habit to implement.

The second way to reduce cross talk includes:

1. Use a continuous return path as the bottom layer.

2. Route all gnd connections from all the components to the bottom plane with as short a trace to the via as practical.

3. Route the gnd via inside the IC footprint if possible. This keeps the path to ground short and does not interfere with signal traces off the IC.

21.6 Fourth Priority: Digital Signals, Congested Signals

Next, route the regions of the board where there is the highest density of traces.

1. Try rotating the ICs or other components to reduce the number of cross-unders required. This should be done during the placement.

2. If you need to use a cross-under, it is better to keep the length of the cross-under short, and use multiple cross-unders in a trace, than to make one long cross-under. This will create a long gap in the return plane and create ground bounce. Adding cross-over return straps will reduce the ground bounce, but shorter gaps will be more effective.

3. Consider routing around components to keep all the trace length on the top surface and visible after components are added.

4. Try to avoid routing under a component. This will make the visual debugging of a trace a little more difficult. If needed, it is better to route under a surface-mount component than to drop a via to the bottom plane and route as a cross-under.

5. Reserve the region under the IC for routing gnd vias and distributing power.

6. Avoid routing traces under the IC if possible. They will be more difficult to see.

7. Be creative in finding routing paths on the top layer so as to avoid using a cross-under.

8. When connecting to a pad, try to enter the pad at 90 degrees so the trace comes directly away from the pad. This just makes routing easier to follow and more esthetically pleasing. There is no manufacturing reason.

9. If there is room, keep the spacing between adjacent signal traces at least 3x the dielectric thickness to the bottom ground plane. This will reduce trace-to-trace cross talk even more. For a board 59 mils thick, this is about 200 mils spacing between adjacent signal lines when practical.

Always design for highest routing density, even if you don't need it. The following are good routing habits:

1. Use the narrowest trace line, providing it will carry less than 1 A average current. This is 6 mils for most low-cost fab shops.

2. Use 20 mil wide traces for power or gnd paths, providing it will carry less than 3 A average.

3. Use the smallest via for the highest routing density. This is 13 mils for most low-cost fab shops.

4. Any component pad connected to a wide trace on the top layer should have a thermal relief on it to thermally isolate it from the plane. This will help soldering.

5. Space the traces as far apart as practical, no closer than the minimum spacing, which is 6 mils for most low-cost fab shops.

6. If you set your PCB grid for 5 mils, the copper traces will route on a 20 mil pitch. If the minimum spacing design constraint is set for 10 mils, the trace-to-trace routing grid will be 30 mils.

7. Rarely will you need a finer routing grid than 5 mils. When you do need to reduce the routing grid, immediately return it to a 5 mil grid.

8. Follow each ghost trace in the rat's nest and make sure they are all routed.

9. Where you need to add a cross-under to the bottom layer, drop a via and change layers. Use both layers for routing the signals, power, and ground connections. Just keep gaps in the ground plane short.

10. Do not use thermal relief vias when changing layers for a signal layer. It takes up too much routing space.

11. If the signal trace is routed over a continuous return plane, it is ok to make the trace long. It is far better to route around a return path discontinuity with a longer line than using a shorter signal line with a screwed-up return path.

21.7 Fifth Priority: Power Paths

1. Once all the signals are routed, route the power paths using 20 mil wide traces. These are safe to carry up to 3 A.

2. If you set the power nets as a directive rule with a width of 20 mils in the schematic, they will all automatically route as 20 mils in the layout. Otherwise, you will have to manually select each power net, then change its width to 20 mils either before or after routing.

3. Consider routing them near the edge of the board outline. This is generally unused routing space.

4. Consider routing the power as a bus line that routes along the edges of your board and then branching off from this bus to reach specific components. The longer paths and higher loop inductance is not important as long as you have local decoupling capacitors.

5. Consider the possibility of adding isolation jumpers in the power distribution paths to isolate a circuit to test different parts. These can act as off-on switches.

6. Use 20 mil wide traces for all power paths. This way, at a glance, you can tell a power path by the wider line width. A 20 mil wide trace can handle 3 A of current with no danger.

7. All voltage regulator modules should have at least one 22 uF MLCC capacitor in close proximity to its output.

8. Every IC power pad should have at least one 22 uF MLCC decoupling capacitor. Preference is to add two of them in close proximity, but only if there is room.

The most important design principle is to route the traces from the IC power pins to the decoupling capacitors and the gnd return in as small a loop inductance as possible. This means:

- ✓ Use short traces from the IC power and gnd traces to the decoupling capacitors.
- ✓ Use wide traces from the IC power and gnd traces to the decoupling capacitors.
- ✓ Keep the power and gnd return paths close together.

The path from the voltage regulator module (VRM) to the decoupling capacitor should be routed with 20 mil wide traces. Its length is not critical as long as there is a decoupling capacitor near the IC's power pad.

21.8 The Silk Screen

The silk screen layer is often called the overlay layer. It is the layer in which you place the printed information that will appear on the top or bottom layer of your board. You should always add some labels to the board in the silk screen or overlay layer, including at least:

- Your name
- Your organization, logo, or website URL
- The board name or id and its rev
- Any special notes

Not all fonts are equally visible at small character heights or pen widths. If visibility is important, use the default font. Otherwise, experiment to determine what your favorite font is for future designs.

Here is a summary of the best design practices for silk screen:

1. The minimum silk screen pen width for most fabs is 6 mils.

2. The typical ratio of character height to pen width should be at least 8 to 1. This means the smallest character height to use is 50 mils. Make it taller so it is easier to read.

3. Add other labels that might help in assembly, test, bring-up, debug, and the user interface.

4. Add component values where possible, in close proximity to the part they identify.

5. Make sure the reference designator labels are near the part they identify.

6. Make sure there is no silk screen printing on top of any soldermask pads.

7. Add polarity indictors outside the footprint of the part.

8. Add pin 1 indicators outside the footprint of the part.

9. Add a label for what each LED indicator is indicating.

10. Add an on or off indictor to each isolation jumper to identify the flag position for open or connected.

11. Think about adding indictors to assist in:

   ✓ Assembling parts
   ✓ Testing and debugging
   ✓ Making it easier for the end user

12. Place the label as close as practical to the component it labels.

13. Label all test points with real names, not just J2, J5. Think about what you want to read when you are testing the board.

14. Identify the signal pin (and what it is measuring) and the gnd pin for testing.

15. Label all header pins if something is to be individually plugged in.

16. Use labels to assist in the assembly operations AND in the final use of the board. Make it easy on the user.

If you are adding silk screen to the bottom of the board, which you should only do *if you have a strong compelling reason to*, note that the artwork on each layer is ALWAYS as it appears looking at the layers from the top of the board. It is always as though you are looking through a transparent board.

This means the silk screen on the bottom layer should be written on the bottom layer as mirror image. If you look through the board from the top to see the silk screen on the bottom layer, you will see the writing on the bottom layer from the back side as a mirror image.

Even though you supply a silk screen layer that is mirror imaged, it will be printed on the bottom layer in the correct orientation.

It is just how it is viewed in the layout tool that it will be mirror imaged.

21.9 Check the Soldermask

The soldermask will cover the entire board on the top layer and the bottom layer. It will prevent solder from sticking to any part of the board underneath it. It will also prevent the copper under the soldermask from shorting any conductors on the surface.

In Altium Designer, the soldermask layer is called the *Solder* layer. A colored region on the Solder layer will have an opening in the soldermask and solder will appear on the board where there is color on the soldermask layer.

The soldermask will be etched away and removed where there is color in the soldermask layer. The soldermask layer is a *negative layer*. The soldermask pads should cover the pads which will be opened to reveal the underlying copper and onto which will be coated the final surface finish.

Note that there is another layer called the *Paste layer*. This would be used to create the *stencil* through which solder paste would be applied to the board. This layer is not needed if you are not generating a stencil. However, if the fab shop is going to do assembly, they will generate a stencil and you will need to provide this layer.

21.10 Soldermask Color

The soldermask that covers most of the board determines the final color of the board. For some applications, you may want a distinctive color. The combination of dark blue soldermask with a gold, ENIG, surface finish and white silk screen is very attractive, for example.

Some companies use a distinctive color as branding. Sparkfun is known for its red-colored boards. Adafruit is known for its purple-colored boards.

When building a prototype that may require extensive debugging, be careful in selecting the soldermask color. Some color soldermasks make it very difficult to see the underlying traces. This means it is all that much harder to debug your board.

For example, a black, blue, or white soldermask is so opaque that it is very difficult to see any surface traces under the soldermask. This makes debugging very difficult.

Do not use these soldermask colors in your first boards where you may want to see the underlying traces. Once your board is working and you are ready for production, you can change the color to anything you want.

In comparison, a green, yellow, or red soldermask shows up the surface traces very easily. This contrast is seen in the examples in **Figure 21.3**.

21.11 Layout — Critical Design Review

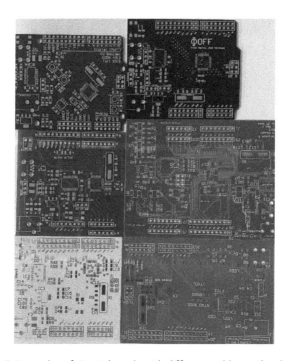

Figure 21.3 Examples of circuit boards with different soldermask colors. When prototyping, try to avoid dark colors such as blue, black, or white. It is difficult to see the traces under the soldermask with these colors.

Keep in mind you can always use one board design for initial bring-up and another board design for production. The bring-up board can have test points, isolation switches, and green soldermask, while the production board can eliminate some of these features and have a black soldermask, if desired.

21.11 Layout — Critical Design Review

Before you release the board to fab, verify it will meet the performance requirements you expect. At the very least, before you present your design for review, check your design for these common mistakes:

1. Indiscriminate copper pour on the signal layer.

2. No ground plane on the bottom.

3. Long gaps in the ground plane without ground straps over gaps.

4. High inductance routing to decoupling caps.

5. No name on your board.

6. Lacking labels for pins.

7. Lacking labeling for parts.

8. Incorrect footprint for pin header sockets.

9. No indicator LEDs.

10. No isolation jumpers.

After you have reviewed the design, another pair of eyes should review your design. More eyes have a higher chance of catching errors. Remember the golden rule in prototype development:

> The earlier in the design process you can catch and fix errors, the lower the cost of your product, the shorter the development cycle, and the higher chance of success of your project.

Always be aware of the time constraints in your schedule.

After you create your Gerber files, you can view them with an independent tool just to verify they are readable. Likewise, when you are evaluating another engineer's design file, you can use a simple, free tool to view each of the layers in the Gerber file.

There are three, simple-to-use, free tools to view Gerber files.

Gerbv is a simple, free popular tool. It can be downloaded here: http://gerbv.sourceforge.net/ .

Cadences' Allegro is a very popular, high-end professional PCB layout tool. They provide a free board file viewer based on their professional tool that will also view Gerber files. It can be downloaded here: https://www.cadence.com/en_US/home/tools/pcb-design-and-analysis/allegro-downloads-start.html . You do have to register, but the download is free.

There is an online Gerber view on the PCBway website: https://www.pcbway.com/project/OnlineGerberViewer.html. Just drag your zip file with the Gerbers and you will see the options to view each layer.

There are some automated Design for Manufacturing (DFM) checker tools all fab shops use. You can take advantage of the DFM tool one on the Advanced Circuits web site.

PCBway has a DFM tool that they run every time you upload a Gerber file.

All these DFM checkers test for is to make sure the features meet the minimum manufacturing constraints. A DFM tool will tell you nothing about:

- Connectivity
- Signal integrity
- Power integrity
- Max current handling
- Cross talk
- Rf design features
- Other noise issues

There are other tools that can also test for electrical performance. An example of a DRC for electrical performance can be evaluated here: https://eda.sw.siemens.com/en-US/pcb/hyperlynx/electrical-design-rule-check/ .

21.12 Practice Questions

1. What are the first parts to place on your board?

2. Why is it a good practice to use through-hole parts for connectors?

3. Why is it important to place decoupling capacitors early in the placement steps?

4. Why should you always keep cross-under lengths to a minimum?

5. After the IC is placed, what is the next component that should be placed?

6. After decoupling capacitors are placed and routed, what is the next trace that should be routed?

Chapter 22
Six Common Misconceptions about Routing

When you read application notes from various vendors or in the popular press, you will encounter at least six guidelines offered to improve performance, but these are misleading and often generate more problems than they solve.

Unless you have a strong compelling reason otherwise, DO NOT follow these six design guidelines:

1. Avoid 90 deg corners.
2. Add copper pour on signal layers.
3. Use 3 different value caps for decoupling.
4. Split ground planes.
5. Use power planes.
6. Use 50 ohm transmission line traces.

22.1   Myth #1: Avoid 90 Deg Corners

Many design guidelines suggest not routing with 90-degree corners in signal paths. There are five reasons often suggested to avoid 90-degree bends:

1. Signal integrity: corners cause reflections and should be avoided.
2. Current crowding: there is a higher current density at the 90-degree bend.

3. A 90-degree bend will hold some of the etch solution on its inside corner due to surface tension. This is called an acid trap.

4. EMI. The electrons accelerate around the edge and radiate more.

5. The sharp edge causes high electric fields that can cause arcing or filament growth.

None of these are compelling reasons to avoid 90-degree ends. Do not avoid 90-degree ends for these reasons.

The signal integrity problems with corners do not arise until above 28 Gbps signals. For applications below 1 GHz, there is no electrical impact from corners. See this article for more information on this problem: https://www.signalintegrityjournal.com/articles/2104-should-you-worry-about-90-degree-bends-in-circuit-board-traces

The increase in resistance from the current crowding at a corner is miniscule and only causes an issue at current densities high enough to heat the trace. You should not use these current levels anyway.

If a 90-degree bend causes an acid trap and a fab vendor is worried, find a new fab vendor. Their rinse process should be good enough to not have to worry about an acid trap.

There is no added radiated emissions from a corner. This is a myth.

Sharp corners do cause higher electric fields, but this should never be an issue in copper traces.

None of these perceived problems with 90-degree bends are compelling reasons to avoid right-angle bends.

The only reason to avoid a 90-degree bend in a routing trace is because two 45-degree bends look neater with cleaner lines than a single 90-degree end. *It is an esthetic issue, not a performance issue.*

If you think two 45-degree bends looks cleaner and more artistically pleasing, use two 45-degree ends instead of a 90-degree bend.

If you don't have room for two 45-degree bends, use a 90-degree bend without worrying about any performance impact.

22.2 Myth #2: Add Copper Pour on Signal Layers

Copper fill or copper pour is extra copper added to a signal layer to fill in all the empty space of the layer between the signal routing. **Figure 22.1** is an example of a close-up of a typical board showing the copper pour between signals on the top layer.

Figure 22.1 An example of a copper pour in the top layer of a Sparkfun Redboard Arduino Uno circuit board. This is not a recommended process.

One important use of a copper pour is to provide a smoother, more uniform inner layer when laminating multilayer circuit boards. This enables a more uniform thickness of the board since the pre-preg resin does not see large areas void of copper that it must fill.

Another important use of copper pour on surface layers is to keep the current density uniform when plating the copper. This keeps the plating thickness the same over the board surface, important for getting a controlled thickness of copper in via holes when a pin must be press fit, or keeping trace thicknesses controlled when a target impedance is required.

Sometimes copper pour is recommended for performance. The various reason offered are:

- ✓ Lowered EMI problems by shielding traces.
- ✓ Lower cross talk between adjacent traces.
- ✓ Better power distribution by using lots of copper.
- ✓ Better overall performance by isolating traces.

All of these performance-related recommendations are absolutely *wrong*. More often than not, a copper fill or pour on the surface layers, unless implemented correctly, will *increase* cross talk and radiated emissions rather than decrease them. Even when implemented correctly, the improvement over not having it is minimal.

A simple example is shown in **Figure 22.2**. Three different simulations are compared. In each case, two 10 mil wide signal traces are separated by 150 mils edge to edge. The far-end cross talk between them, one metric of cross talk, is calculated. One example is what would appear if there were no copper pour. When copper pour is added, and the ends are left floating, resonant coupling would appear, dramatically increasing the cross talk at certain frequencies. Finally, a copper pour is added but with many shorting vias.

22.2 Myth #2: Add Copper Pour on Signal Layers

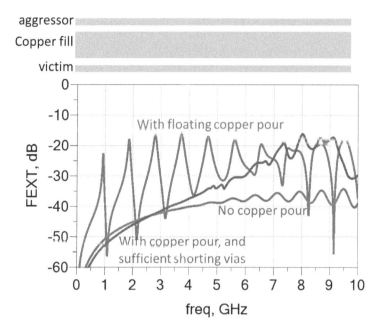

Figure 22.2 A simulation of the cross talk between the aggressor and victim under three conditions of copper pour: no pour, floating pour, and copper pour with many shorting vias. Simulation performed by Saish Sawant.

The risk of adding a copper pour to the surface of your board is increasing the cross talk, due to resonant coupling, over what would appear if there were NO copper pour. Even in the best case, if sufficient shorting vias were added, the benefit of adding a copper pour is almost negligible. The cross talk when the spacing is large is already very low. There is no compelling advantage of adding a copper pour but a very large risk.

Sometimes copper pour is recommended for power. Remember that even a 20 mil wide trace provides perfectly adequate routing for as much as 3 A in many cases.

If you add a copper pour and connect it to the power net, it is difficult to trace the routing by eye. You will have a harder time debugging your board. And the series resistance will be limited to the regions in the power routing where the pour doesn't fit or there is a neck down. This is very hard to spot.

There is no compelling performance advantage to ever use a copper pour, contrary to what is often recommended. It will solve no problem and sometimes make it harder to debug your board. There is the potential for more problems, such as isolated (orphaned) copper islands and added crosstalk with a copper pour on a signal layer compared with not using it.

Just throwing copper on the board wherever there is open space solves no problem.

It may be important to add copper to outer layers to balance the copper plating currents to get a uniform copper thickness. But don't use copper pour. The fab vendor will oftentimes add thieving pads, if requested. See this article by Lee Ritchey: https://www.signalintegrityjournal.com/articles/649-thieving-in-printed-circuit-boards

An example of a copper surface on an outer layer with circular thieving pads to balance the plating thickness of copper over the whole surface is shown in **Figure 22.3**. This sort of copper fill will have no impact on performance but will aid in manufacturing.

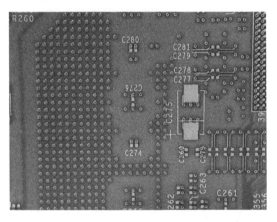

Figure 22.3 An example of using small, isolated islands of copper pads on the surface to balance the platting current density during the manufacturing process.

22.3 Myth #3: Use Different Value Decoupling Capacitors

This recommendation is that for every power pin on an IC, add a 10 uF, a 1 uF, and a 0.1 uF capacitor. The rationale is that the 0.1 uF capacitor is a high-frequency capacitor and the 10 uF capacitor is a low-frequency capacitor and just provides capacitance at low frequency.

This is a correct statement when dealing with through-hole, leaded capacitors. However, with SMT MLCC capacitors, a 10 uF capacitor is just as good a high-frequency capacitor as a 0.1 uF capacitor. This design guideline no longer applies to SMT capacitors.

Instead, it is far more important to use low mounting inductance for the capacitors. If you are not going to do your own analysis, then as a general design guideline, use at least one 10 uF capacitor.

For more details on the myth of three capacitors, check out this article in *SI Journal*:
https://www.signalintegrityjournal.com/articles/1589-the-myth-of-three-capacitor-values

22.4 Myth #4: Split Ground Planes

It is commonly recommended to split ground planes to prevent the noise from one ground getting on another ground.

There is only one case in which this might happen and that is when noise on the order of 1 mV or less, below 100 kHz bandwidths is important and when there are large, low-frequency currents flowing in the ground plane, usually from high current (> 10 A) power rails.

For signal frequency components higher than 100 kHz, return currents will be flowing in the ground plane directly underneath the signal (or power) current. The IR drop in the ground planes will be localized as underneath the signal path and not spread out under the plane.

If this 1 mV of low frequency noise is important in your application, such as when you are using a 24-bit ADC or measuring a low-level analog signal, you should use a differential amplifier. This will eliminate any impact from the IR drop in the ground plane.

22.5   Myth #5: Use Power Planes

Many design guidelines will suggest in a 4-layer board to use the inner two layers as power and ground planes. This is to provide high-frequency capacitance to the ICs, which helps reduce EMI and high-frequency noise from the power pins of the IC. This is totally misleading.

The "high-frequency" capacitance really comes from the capacitance that is already on the die. Generally, there is far more capacitance on-die than can be added in the power and ground planes in the circuit board. And the package lead inductance limits the impedance looking into the planes at the high-frequency end.

The requirement for a power plane is only if there is very high DC current to distribute. After all, even a 20 mil wide trace will handle 3 A easily. A 100 mil wide trace will handle 10 A. These traces can be routed on signal layers. If you have more than 20 A of current that has to be routed to multiple devices on a board, a plane may be a way to do this.

As with every design guideline, you always have to do your own analysis. Generally, in a 2-layer, 4-layer, and oftentimes a 6-layer board, there is rarely a need for a power plane in the board.

Remember three important PDN principles:

1. A 20-mil wide trace can carry 3 A of DC current with no problem. A 100-mil wide trace can carry 10 A of DC current.

2. A low inductance in the power delivery path is most important between the IC power pins and the decoupling capacitors. This is where the wide conductors in the power path are needed.

3. After designing for connectivity, interconnect design is all about designing for lower noise.

The most important source of noise to manage is ground bounce caused by discontinuities in the return path. The lowest noise will be with a continuous ground plane on the adjacent layer to the signals.

In a 2-layer board, the bottom layer should be ground. This leaves the top surface for power routing. Use power puddles as short, wide polygons between the decoupling capacitors and the IC power pads. This is where low inductance is required.

Any power path not between the decoupling capacitors and the IC power leads on the IC serves no purpose except delivering DC current.

In a 4-layer board, it is more important to use the two inner layers as ground planes so they can be the return path for signals on the top and bottom layers.

This way, when a signal changes layers and connects with a via between layer 1 and layer 4, the return current will need a via to connect the return current from layer 2 to layer 3. Wherever you use a signal via, add an adjacent return via to provide a low impedance path for the return current to flow from the first ground plane to the second.

If the layer 3 is a power plane, you cannot add a return via. Route the power as wide traces on layers 1 and layer 4. Likewise, add a return via whenever the power changes layers.

If you are designing a board with more than 20 A of current, you should use either wide power paths on the signal layers or consider a power plane.

22.6 Myth #6: Use 50 Ohm Impedance Traces

All interconnects are always transmission lines. Reflections will always happen when the instantaneous impedance the signal sees changes.

For example, there will be a reflection when the signal sees some instantaneous impedance in the trace on the board and then the open at the receiver, RX. This reflection will head back to the transmitter, TX.

When the TX has a low output impedance, such as a fast CMOS driver, multiple reflections between the low impedance of the TX and the high impedance of the RX can create what appears at the RX as ringing noise. **Figure 22.4** shows an example of the measured ringing from reflections back and forth between the RX and the TX.

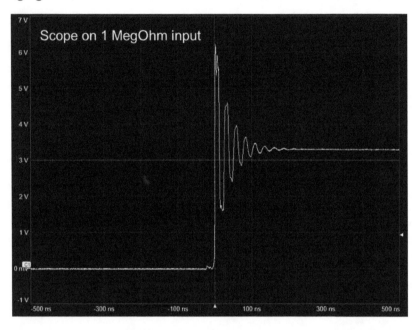

Figure 22.4 Measured ringing noise from reflections between the 1 Meg input to a scope and the low impedance of the TX.

The ringing period is 4 x TD of the transmission line. The number of cycles it lasts is related to the impedance difference between the TX and the interconnect. Generally, for a typical CMOS driver with an output impedance about 20 ohms, it will last for 1 to 3 cycles.

This behavior will always happen when the TX is a low impedance and the RX is a high impedance. But there are three conditions that might make this ringing noise insignificant and nothing to worry about:

Condition 1: If the output impedance of the TX is close to the interconnect's characteristic impedance, there will be no reflection from this interface and no ringing noise. This is actually what we try to engineer in a source series termination strategy.

Condition 2: If the rise time of the signal from the TX is about 4x the TD of the interconnects, the reflections will happen, but they will be smoared out during the rising edge and will not be apparent. **Figure 22.5** shows an example of the reduction in the ringing amplitude at the RX as the rise time, RT, of the signal increases, compared to the time delay, TD, of the interconnect.

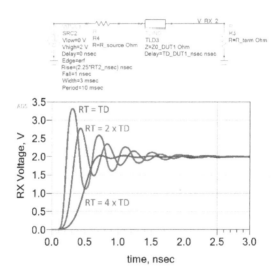

Figure 22.5 Top: example of a simple circuit with a signal source, a source impedance, driving a transmission line with a high impedance receiver. Bottom: the simulated voltage at the receiver as the rise time of the signal increases compared to the time delay of the transmission line. Simulated with Keysight's ADS.

The wiring delay of a typical interconnect on a circuit board is 0.17 nsec/inch. A trace that is 3 inches long will have a time delay of about 0.5 nsec. If the rise time of a signal is 4 x 0.5 nsec = 2 nsec or longer, the transmission line effects of the traces on the board may not influence signal quality.

Condition 3: If the ringing noise happens during a time when any RX is not looking at the input, for example when the ringing noise is outside the sample and hold time of the RX, then ringing noise may not affect the RX response.

For most small boards using uC devices, the reflection noise from transmission lines will not generally cause a problem that requires controlled impedance interconnects and a routing or termination solution.

However, if your traces are longer, or your rise times are shorter, if output impedances of TX are low and clock periods short, then transmission line effects will be important, and you should design all traces as controlled impedance interconnects at a target impedance. In this case, a termination strategy and routing strategy are very important.

These sorts of boards, where transmission line effects matter, are referred to as "high-speed" or "controlled impedance" boards.

These topics are covered in other textbooks on signal integrity and transmission lines.

22.7 Practice Questions

1. Does a 90 deg bend cause a reflection?
2. Do electrons accelerating around a 90 deg bend radiate more?
3. Should you avoid a 90 deg bend?
4. What is a good reason to route bends at 45 deg?
5. What is the potential risk of adding a copper pour?
6. Why do some designers recommend adding a copper pour?
7. How should you route power from the VRM to devices?

8. What is the recommended practice for adding decoupling capacitors to an IC?

9. What is the old motivation for using three capacitor values and why is this no longer a valid reason?

10. What is the only condition to consider using a split ground plane?

Chapter 23
Four-Layer Boards

When we run out of room to route the traces in a 2-layer board, and we've used as many cross-unders as practical, it is time to consider using a 4-layer board.

Unless you are using an assembly shop that can assemble on both sides of the board, you should plan to place all the parts on the top side of the board. This leaves the top layer to route whatever traces as will fit and to route to vias that connect to the other routing layers for more routing.

With a 4-layer board, it is important to consider the stack-up options and select the optimum stack-up, not just for connectivity but also for performance.

23.1 Two-Layer Stack-Ups

In low interconnect density applications, we use two layers in the board stack-up. If we are just designing for connectivity and the interconnects are completely transparent, we use the top layer for component attach and route signals indiscriminatingly on either of the two layers.

Connecting the traces on the top layer to the bottom layer are vias. These are holes drilled through the board, cutting through traces on the top and bottom layers, exposing the edges of the trace inside the hole. Through a two-step process, a thin layer of copper is electroless plated inside the hole, connecting the exposed metal on the top layer with the exposed metal in the bottom layer. Then the thin seed layer of copper is electroplated to a final thickness of typically 1 oz copper. These vias are called plated through-hole (PTH) vias.

An example of routing in a 2-layer board with PTH vias is shown in **Figure 23.1**. In this example, routing is on both layers. Typical board thicknesses are 64 mils, 32 mils, or 16 mils.

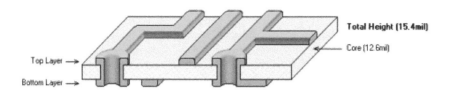

Figure 23.1 Example of a 2-layer board with routing on both layers. Courtesy of Altium.

Generally, this board is constructed from a single layer of fully cured dielectric, referred to as a core, with copper foil laminated to both sides, patterned, and etched away.

If we design just for connectivity, we run the risk of two important sources of switching noise: signal-to-signal cross talk and power rail noise.

Higher cross talk noise between signal lines can result from the high mutual inductance between signal-return path loops. The way to minimize this problem is to use a continuous return plane for the signals routed as a solid ground plane on the bottom layer.

This way, the top layer becomes mixed signal and power routing, with the bottom layer reserved for the return path. Where needed, the bottom layer can be used for cross-unders. To reduce their impact on cross talk, keep the length of the cross-unders short and add return straps if signals cross over the gap in the return plane.

The second switching noise problem, power rail noise, can be reduced by using local decoupling capacitors close to the IC with low loop inductance between the IC power pin and the capacitor.

When we require higher interconnect density than can fit on the top layer with some cross-unders, it's time to consider switching to a 4-layer board.

23.2 A 4-Layer Board

A 4-layer board has four metal layers. It is usually constructed from a 2-layer board, built as a core with glue or pre-preg layers added to each side with a copper foil laminated on top and bottom. An example of a 4-layer construction used by PCBway.com is shown in **Figure 23.2**.

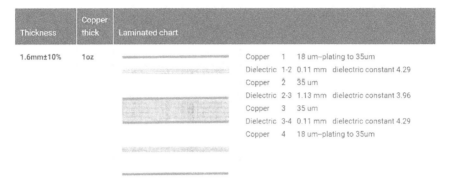

Thickness	Copper thick	Laminated chart		
1.6mm±10%	1oz		Copper 1	18 um–plating to 35um
			Dielectric 1-2	0.11 mm dielectric constant 4.29
			Copper 2	35 um
			Dielectric 2-3	1.13 mm dielectric constant 3.96
			Copper 3	35 um
			Dielectric 3-4	0.11 mm dielectric constant 4.29
			Copper 4	18 um–plating to 35um

Figure 23.2 PCBway construction for their low-cost 1.6 mm thick 4-layer board. Note the thicker core, with thinner dielectric between the top and bottom layers and the core.

Another example is shown in **Figure 23.3** from another low-cost vendor, JLCpcb.com. In this stackup, the dielectric layers in the pre-preg layers, composed of 7628 type glass weave, are 0.2 mm or 8 mils thick.

Layer	Material Type	Thickness	
Top Layer1	Copper	0.035 mm	
Prepreg	7628*1	0.2 mm	
Inner Layer2	Copper	0.0175 mm	
Core	Core	1.065 mm	1.1mm (with copper core)
Inner Layer3	Copper	0.0175 mm	
Prepreg	7628*1	0.2 mm	
Bottom Layer4	Copper	0.035 mm	

Figure 23.3 Stack-up for a 4-layer board offered by JLCpcb.com using a pre-preg layers composed of 1-ply of 7628 type glass.

The copper on each side of the core is patterned before the sandwich is laminated. After the output two layers are laminated to the core, vias are drilled through the board and plated with copper. Then the top and bottom layers of copper foil are patterned, soldermask applied and patterned, and the surface finish applied.

23.3 Four-Layer Stack-Up Options

In a 4-layer board, we have options of using each layer for whatever purpose we choose. To the PCB fab shop, each layer is just patterned copper.

What is the best selection for each layer? This is part of the layer stack-up decisions.

A 4-layer board is always more expensive than a 2-layer board. The price can increase by 10x depending on the fab vendor. At PCBway, a minimum order of 2-layer boards is $5 each. A minimum order of 4-layer boards is $50 each. But at JLCpcb, a minimum order of 2-layer boards is $2. A minimum order of 4-layer boards is only $7.

We only use four layers if we have to. This is generally if we cannot fit all the routing we need to do on a 2-layer board.

Unless you have a strong compelling reason, don't use four layers. It's the strong compelling reason that drives how we select the layers.

We've seen that a 20 mil wide trace can easily handle 3 A of current. If your application requires less than 3 A of current, you do not have to route power in a plane. You can do just as well continuing to use 20 mil wide traces.

If your application requires 20, 30, even 50 A of current, you need to use wide traces to distribute the current and this may require at least one of the layers, maybe even two of them, to be dedicated to carrying the DC current.

For such high current applications, you also might want to consider using thicker copper than 1 oz for its lower resistance and higher current-carrying ability.

Usually, the compelling reason to switch to four layers is routing density. If we can't grow the size of the board but are limited in a specific form factor, and we have a lot of traces we have to route that we cannot fit in 1 or 1 ½ layers, switching to four layers is the right answer.

How shall we select the layer assignment for a 4-layer board? **Figure 23.4** shows six of the most common selections for layer assignment in a 4-layer board. They each have their own plusses and minuses based on the conditions of:

- ✓ Routing density
- ✓ Good signal/power integrity and low noise
- ✓ Good high current-handling capability
- ✓ Ease of assembly, test, debug
- ✓ Mechanical shape and rigidness

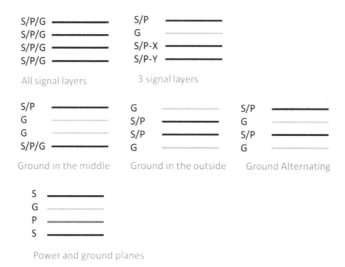

Figure 23.4 The six options for layer assignment in a 4-layer board.

If connectivity is the only thing that is important and high routing density is the driving force, and there are no concerns for signal integrity, power integrity, or electrical noise, then using *all layers as signal layers* is an option. This should be the least-preferred option, as noise will often be an issue when any digital signals switch states.

As we have seen, the most important way of controlling the noise generated when signals switch is using a continuous return plane, usually labeled as ground. If interconnect density is an important driving force, and good signal density is important, the *3-signal layer* option might be acceptable.

In this configuration, the top layer 1 has components, signal routing, and power routing. Layer 2, the ground plane, is continuous except for clearance holes for the vias.

Layer 3 is a signal and power layer with traces all going in the X-direction. Layer 4 is a signal and power layer with all traces going in the Y-direction. This way, the routing is neat and orderly and there are no crosses on the same layer.

The signals on layer 3 use the ground plane as their return path, and the signals on layer 4 use the same plane as their return. However, since the signal traces are routed orthogonally, the return currents for the X and Y layers do not overlap and there will be no ground bounce between signal traces on these layers.

The most important condition is to make sure that the signal vias do not have overlapping clearance holes in the ground plane. This would create inadvertent gaps in the return plane and can contribute to ground bounce.

23.4 Stack-Up Options with Two Planes

If you only need two signal layers for routing, the middle three options are candidates. They each have two ground planes. The advantage of two similar voltage-planes is to enable the connection of return currents between planes when signals change return layers.

For example, a signal on the top layer, layer 1, will have its return current in the plane on layer 2. If a via connects this signal to layer 4, the return current for this trace would be on the adjacent plane, on layer 3.

The signal current passes through the via from layer 1 to layer 4. We need to provide a low impedance path for the return current to pass from the plane on layer 2 to the plane on layer 3.

The lowest impedance path is a direct via connection. We can only connect layer 2 to layer 3 with a via if the planes have the same voltage. This is why using two ground planes is preferred. It enables the use of a shorting via between the two planes in proximity to every signal via. This dramatically reduces the cross talk when a signal transitions between layers.

Of course, the shorting via between the planes would also be a PTH passing through the entire board. This is the lowest-cost via.

When the middle two layers are ground planes, add a return via between the two planes adjacent to any signal via that passes from the top layer to the bottom layer.

Many PCB designs recommend using *power and ground planes* in the stack-up for a 4-layer board. If the currents used by components exceeds about 20 A, then the power plane may be the right path to take to provide a low IR drop distribution for power.

But, if the currents are less than 20 A, they can be distributed using 20-400 mil wide traces. The danger of using a separate power and ground plane is that the traces on the top layer have return currents in the ground plane but traces on the bottom have their return currents in the power planes.

When a signal transitions between the top layer and the bottom layers, the return currents will have to flow between the two planes. If there is no low-impedance DC path, there will be some impedance between these layers. The return currents flowing through this impedance will generate ground bounce noise.

To reduce the ground bounce noise when signal layers transition, you need to engineer a low-impedance path for the return current.

You cannot add a via between these planes since that would short power and ground.

The only option is to use a DC blocking capacitor connecting vias between these two planes. The loop inductance of the DC blocking capacitor will be 5x to 10x higher than for a direct via connection, which means the impedance will be 5x to 10x higher. It may work, but it is a riskier approach.

For example, the loop inductance of a 1206 DC blocking capacitor might be 5 nH. If the dI/dt of one signal is 30 mA/nsec, the ground bounce noise across the 5 nH inductance would be 5 nH x 30 mA/nsec = 150 mV for one signal switching. With five signals switching through this same DC blocking capacitor, the switching noise could be as large as 750 mV.

To avoid this potential source of switching noise, make the two internal planes both ground planes so they can be connected with a via. And place a return via in proximity to each signal via.

If you do not have a strong compelling reason to use a power plane, don't. Route the power using wider traces and use local power puddles between the IC and their decoupling capacitors.

23.5 The Recommended 4-Layer Stack-Up

Which of the three approaches using two signal layers and two ground layers is preferred?

The ground in the middle stack-up is a good default. Unless you have a strong compelling reason otherwise, use this stack-up. It provides the low noise and two layers of routing.

The components are on the top layer. You can see the routing on the top and bottom layers and use this as a debug verification.

It is also symmetrical so will be insensitive to warping.

The stack-up with the ground on the outer two layers and signals on the inner two layers has its own set of problems. The top layer will be the component attach layer and a copper fill for the top ground layer. Due to the tight space between components, there may be large gaps in the top ground plane, resulting in some ground bounce for signals on layer 2.

It is also difficult to see the routing to verify the design.

The ground alternating stack-up is an option. The top layer is component-attach and some signal and power routing. It has a solid ground underneath it.

The other routing layer is between two ground planes. Many app notes suggest that using this stack-up of a signal with planes above and below, referred to as *stripline*, is a lower-noise geometry and will be more likely to pass an EMC certification test.

This is generally not true. It is possible for microstrip traces to pass an FCC test just as easily as a stripline. The key element is continuous return paths. Signal traces crossing gaps in the return path are the real cause of FCC EMC certification failures. This can happen just as easily in stripline traces as in microstrip traces if the return planes are not continuous.

This stack-up is also slightly asymmetrical. It has the potential of warping under thermal expansion.

23.6 When Signals Change Return Planes

In a 4-layer stack-up with the inner two layers as ground, the signal's return current still has to transition from one plane to the other when there is a signal via. With no adjacent return via, the return current will flow through the impedance of the cavity composed of the two planes. This is illustrated in **Figure 23.5**.

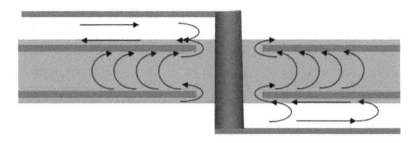

Figure 23.5 Return current when a signal changes return layers.

As the signal passes through the cavity, the return current moves from the top surface of the top layer to the inside surface by flowing around the lip of the clearance hole. Once inside the cavity, the return current flows through the radial transmission line of the cavity and propagates to the edge of the board where it reflects. These multiple reflections generate resonant ringing noise inside the cavity.

The resonant frequency of a cavity composed of two planes with FR4 between them is about

$$f_{res}[\text{GHz}] = \frac{3}{\text{Len}[\text{in}]}$$

If the longest length of the cavity is 3 inches, the resonant frequency of the cavity will be 1 GHz. If the bandwidth of the signals is << 1 GHz, there will be no signal components to excite the cavity and little cavity noise.

A 1 nsec signal will have a bandwidth of about 0.35/1 nsec = 350 MHz. It would take a board, 10 inches on a side, to have a resonant frequency of 350 MHz where the signal could excite cavity noise and be a problem.

This problem of cavity noise cross talk is really an issue only for signals with rise times shorter than 1 nsec. While in principle, signals switching return planes will inject noise in the planes, in practice, the noise will only be a problem when the signal rise times are very short, 1 nsec or shorter.

When signal rise times are 10 nsec or longer, there will be little noise injected in the plane cavities. However, it is still a good habit to design for controlling the return currents when signals switch return planes. It will always reduce the radiated emissions from the low level noise in the planes.

To reduce the potential of via-to-via noise, add a return via that connects the return current between the two planes, in close proximity to each signal via, as shown in **Figure 23.6**.

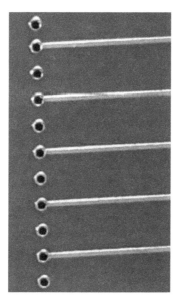

Figure 23.6 An example of a collection of signal vias transitioning from the top layer to the bottom layer of a 4-layer board with adjacent return vias in close proximity.

23.7 Practice Questions

1. What is the purpose of a via?

2. What does it mean for the stack-up to be asymmetrical?

3. What is the potential problem if the stack-up is asymmetrical?

4. Where is a buried via located?

5. Why should you avoid using a blind or buried via?

6. What is the preferred stack-up for a 4-layer board? Why?

7. How does the return current change return planes when a signal via transitions from the top to the bottom layers?

8. If the two inner planes are power and ground, what is the danger of adding a return via?

9. What do the two inner planes have to be in order to add a return via?

10. What is the preferred type of via to use to connect the two inner ground planes?

Chapter 24
Release the Board to the Fab Shop

After your design has passed your CDR, you export or output the Gerber files and the NC drill file. These should be added to a single folder and zipped. This is the design file that will go to the fab vendor.

24.1 Gerber Files

Each EDA tool uses a slightly different extension for the Gerber files. In Altium Designer, for example, the Gerber files use extensions to identify the content:

- .GTL is the Top Layer metal
- .GTO is the Top Overlay, which is the silk screen layer on the top layer
- .GTP is the Top Solder Paste stencil screen openings for screen printing solder paste
- .GTS is the Top Soldermask openings through which the solder on the pads will show on the top layer
- The label of B instead of T, as in .GBL, means the Bottom Layer metal
- There will also be a keepout layer or board outline layer

After you create the Gerber files, export the NC drill file. This is a text file that contains the coordinate and size information for the NC drill machine for each drilled hole. This is the information about all your vias and through-holes. After the Gerber files are prepared, the next step is to submit them to the fab house for review.

The final zip file which will be sent to the fab shop should have the Gerber files and the NC drill file. Do not add all the files possible, only those which are needed by the fab shop. Sometimes, if you include mechanical layers, these layers get printed as copper layers and may short out pads or traces. If the fab shop does not explicitly need to use a mechanical layer in fabricating your board, do not include it. This will reduce the risk of an unwelcome surprise.

An example of the specific files you will see in the zip file sent to the fab shop is shown in **Figure 24.1**.

Analog Switches.apr	CAMtastic Aperture Data
Analog Switches.DRR	Altium NC Drill Report File
Analog Switches.EXTREP	EXTREP File
Analog Switches.GBL	CAMtastic Bottom Layer Gerber Data
Analog Switches.GBO	CAMtastic Bottom Overlay Gerber Data
Analog Switches.GBS	CAMtastic Bottom Solder Mask Gerber Data
Analog Switches.GKO	CAMtastic Keepout Layer Gerber Data
Analog Switches.GTL	CAMtastic Top Layer Gerber Data
Analog Switches.GTO	CAMtastic Top Overlay Gerber Data
Analog Switches.GTP	CAMtastic Top Paste Mask Gerber Data
Analog Switches.GTS	CAMtastic Top Solder Mask Gerber Data

Figure 24.1 An example of the files inside the zip folder that would be sent to the fab shop for a 2-layer board.

If the fab shop is also doing the assembly, you will need to provide the BOM in spreadsheet form and the pick-and-place file. These three files are the design files submitted to the fab shop.

24.2 Cost Adders

Unless you have a strong compelling reason, always use the default features for your board ship to keep the cost to a minimum.

If you are paying more for your board than the minimum, you should be aware of the value you are gaining.

For example, the lowest-cost conditions for PCBway are:

- ✓ No finer than 6 mil wide trace and 6 mil wide space
- ✓ No smaller than 13 mil drilled hole
- ✓ No smaller than 25 mil diameter capture pads with 6 mil annuluses
- ✓ Two layers
- ✓ Board dimensions less than 100 mm x 100 mm or 3.9 inch x 3.9 inches
- ✓ If there is no cost adder for a quantity of 10 boards, order 10 boards.
- ✓ Thickness of 1.6 or 0.8 mm
- ✓ No finer than 6 mil wide pen for the silk screen
- ✓ Using FR4 and the default temperature range
- ✓ Surface finish of HASL leaded

The soldermask color and silk screen color are no extra charge.

Be mindful of your selections so you do not add unnecessary cost to your board.

For example, unless you really need it, do not:

- ✓ Ask for ENIG or bare copper
- ✓ Use a narrower feature than 6 mils
- ✓ Use four layers
- ✓ Ask for buried vias
- ✓ Use a 0.4 mm thick board
- ✓ Ask for HASL lead-free
- ✓ Specify a board larger than 3.9 inches on a side

24.3 Board Release Checklist

- ✓ You have reviewed the schematic with another pair of eyes
- ✓ You have reviewed the layout with another pair of eyes
- ✓ Board has your name, board name and date code on the board
- ✓ You have added meaningful labels to all indicator lights
- ✓ You have added meaningful labels to all switches
- ✓ You have added meaningful labels to all test points
- ✓ Your silk screen pen width (stroke) is larger than 6 mils
- ✓ Your silk screen character height is at least 50 mils tall whenever possible
- ✓ All features used are within no-cost-adder parameters unless you are willing to pay the extra price
- ✓ You have exported the Gerbers and the NC drill files
- ✓ You have the outline in the Gerber files
- ✓ You have verified the Gerbers are acceptable and cost is acceptable
- ✓ You have three files, if asking for assembly:
  - Gerbers
  - Pick and place
  - BOM
- ✓ File names:
  - Have no spaces
  - Start with your last name
  - Use same last name for all three files
  - Have file type, Gerber, PnP, or BOM
  - Have board name
  - Have a date code

- All separated by "_"

24.4 Practice Questions

1. In the Gerber file, what does GTS mean?
2. What does GTL and GBL mean?
3. Where do the NC drill files go?
4. What should be in the file name?
5. What is the extension name of the keepout layer and why is this important to add?
6. Why should you avoid including unnecessary mechanical layers in the Gerber file?
7. What are three important features to check in your layout before creating the Gerber files?

Chapter 25
Step 6: Bring-Up

Our goal in designing a prototype is to get it right the first time.

In the ideal world, you design the widget, acquire the parts, assemble them, and the product works exactly as expected when you turn it on. Every performance spec feature is tested, and the measured behavior meets or exceeds the spec.

Of course, our goal in developing a robust design and manufacturing process is to attain this ideal.

But sometimes your prototype does not work the first time. When it doesn't, we switch modes to find and fix the problem. Our goal is to get it right the second time as quickly as possible.

25.1 Does Your Widget Work?

The term "work" is defined based on the expectation of the product as defined in the specification or the POR. This is just the starting place to identify catastrophic or pathological or gross failures.

For example, performance that could be demonstrated for the prototype to "work" might be:

- ✓ The power lights go on.
- ✓ The total current draw from the power supply is less than some high limit.
- ✓ If a processor of some sort is incorporated, the widget boots up and communicates with a host.
- ✓ Some set of specific commands can be executed and the responses match what is expected.

✓ A selection of stimuli are input to nodes in the circuit and the responses on other nodes match the acceptable values based on the specification of the product.

The purpose of the bring-up phase is to verify the prototype meets as many of the criteria of "working" as possible.

The output of the bring-up phase is to have a prototype that meets these preliminary criteria so that the next, more thorough phase of testing and characterizing can be completed. When all the possible tests have been performed and the prototype passes all of them, it can move to the next phase of evaluation, including margin and reliability testing.

In this next phase we do characterization, measure performance metrics, extract figures of merit that describe the performance, and evaluate the performance margin.

In this last phase, we also explore options to optimize the cost-performance of the design. This can be either to pay a little more for more performance (more bang for the buck) or see if we can reduce the cost at no or little sacrifice in performance (cost reductions).

25.2 Prototype or Production Testing

There is a difference between testing a prototype for the first time and production testing.

In production testing, the design and component selection has already been verified. Production testing is really about testing for component defects and assembly defects.

What makes testing in the prototype phase more difficult than in production is that not only are you testing for component or assembly defects, but also testing for fundamental circuit design and layout problems.

25.2 Prototype or Production Testing

The difference between a defect problem and a design problem is that a defect problem arises in some units but not all, while a design problem is inherent in EVERY unit produced.

Distinguishing what is a design problem or a component or assembly defect problem is often very challenging.

As mentioned earlier, you want to do everything you can to increase your chance of catching design problems before they are introduced into your circuit board prototype, such as:

- ✓ Simulate what you can.
- ✓ Build a solderless breadboard prototype to test the circuit.
- ✓ Acquire an eval board to get familiar with the signature of signal responses.
- ✓ Test your code on a similar commercial module.
- ✓ Build a simple eval board of an important functional block to test this part of the circuit.

The process for testing a prototype to determine if it works consists of six important steps:

1. Create specific tests to verify if the product works.//
2. If a test fails, find the specific problems that are preventing the prototype from working.
3. Identify their root cause.
4. Fix any assembly or component defect problems.
5. Fix any schematic design or component selection problems or layout design problems by replacing components, cutting traces, or adding engineering change wires.
6. What can't be manually fixed in the prototype unit, respin the design and reorder the board and start over again.

Obviously, if the problem can't be fixed in the first prototype units and a respin of the design, layout and circuit board is required, there will be a delay of at least 1-2 weeks and the additional expense of a new board run.

The earlier in the design process you can find and fix potential problems, the lower the total cost and the shorter the development time. Extra effort invested in the plan of record, early design evaluation, the preliminary design review, and the critical design review may be an excellent return on investment.

Before a new board is redesigned and respun, it is important to have confidence you have found the correct root cause of all the problems and they will be fixed in the respin.

25.3   Design for Bring-Up

The goal in *designing for bring-up* (DFB) is to use best design practices which will help us find and fix any problems so we can get to a verified, functioning prototype that meets our performance specs as quickly and efficiently as possible. This is the bring-up process.

Keep in mind that the board you use for bring-up and test does not have to be your final production board. Oftentimes, we will add features to the initial board to aid in bring-up and debug that, once the design and component selections are verified, we can remove.

In an *ideal world*, the design will be perfect, the board will come back from the fab shop flawless, and the assembly process will be perfect. When you power up your board, it will work exactly as expected under all conditions.

Unless you are skilled in the art and have built many boards knowing how to design like Ralphie's mom and have confidence your design is robust and will work the first time, *assume it will not work the first time*.

If you assume the board will not work after assembly, you take the approach of testing the board as you assemble it, looking for the problems.

It is far easier to find a problem and fix it as you assemble than if the entire board is assembled and then turned on for the first time.

Get in the habit of assembling a few parts at a time and testing as you go rather than assembling the entire board and then turning it on.

If you get to the point where the section is assembled, and you do not see a problem and it works, you can move to the next test phase.

The goal in bring-up is to follow a strategy that will help you identify the problems and find their root cause.

If your prototype is already assembled, you can use isolation switches to selectively turn off sections of your circuit so you can isolate problems and debug specific circuits one at a time.

25.4 Find the Root Cause

The most important step in fixing a problem is to find the root cause of the problem and then fix the problem at the source, either as a redesign or as a repair of the current board.

The fastest and most efficient way of fixing a problem is to first identify the problem and then find the root cause.

If you have the wrong root cause for a problem, your chance of fixing the problem is based purely on luck.

To illustrate the importance of having the correct root cause, **Figure 25.1** shows the consequence of identifying the wrong root cause.

Figure 25.1 The consequence of having the wrong root cause.

When your board is not working, resist the temptation to try changing random things until it works. This may appear to make the problem go away, but you will have no idea what the root cause was or even if you really fixed it.

When your board is not working, just watch. Measure what you can, such as all the pins of the suspect IC. Look for clues that hint at the root cause. Measuring the voltage and current draw on the power rails is a powerful diagnostic of the root cause of potential problems.

After you have identified a possible root cause, then you can test for it.

Are the pins connected correctly? Review the pin diagram in the datasheet and carefully check the routing of each pin on the board.

Are you really using the part in the correct way as described in the datasheet or reference design?

Is the device getting power? Check the power connections, then look at the voltage on the power pin with an oscilloscope. Avoid using a DMM to measure a voltage as excessive noise on the power rail may look like a DC voltage with a DMM, but in reality, may be very high-frequency noise. A scope is always the preferred tool to measure a voltage on any node as it will reveal potentially more clues than just a DMM.

25.5 Problems to Expect

In the real world, there is always the potential of a problem introduced into your prototype. While there are many, many ways of screwing up a product so it doesn't work the first time, there are a handful of common problems, related to each phase of the project.

This list of the possible sources of problems can be used as a checklist to review at each phase of the project.

1. Problem in the *conceptual design*: Even if your plan is executed perfectly, the plan may be flawed. You wanted to use a haptic vibration source to modulate the intensity of a laser. But there was no coupling between the vibrations and the laser. The laser intensity did not change.

2. Problem in the *part selection*: You chose the wrong part. The design is implemented perfectly, but the BOM is incorrect. You needed a 1 microsecond rise time on the timer. You chose a 100 kHz bandwidth part rather than the 10 MHz bandwidth part. You did not read the datasheet very well and did not specify the correct part.

3. Problem in the *schematic*: You wired the parts up incorrectly. The layout was executed exactly as per the schematic, but the schematic was wrong. The TX of one part should have gone to the RX of the other part, not to the TX.

4. Problem in the *connectivity* of the layout: When you routed the trace on the layout from pin U1.4 to U4.9, the trace did not make contact with the pad to complete the circuit.

5. *Excessive noise* picked up due to a signal or power integrity layout issue: You routed a trace over a gap in the return path, creating too much switching noise that caused a false trigger.

6. *Wrong part* assembled to the board: You assembled a 22 pF capacitor to the LDO output instead of a 22 uF capacitor.

7. *Missing part*: You forgot to add the decoupling capacitor to the LDO output.

8. *Bad part* assembled to the board: Your TMP36 sensors were all ESD damaged before you got them, or thermally damaged by your too-hot soldering iron, or the batch of parts you purchased were all bad parts.

9. *Manufacturing defect* in the fab: The trace between two pads has a defect, causing it to be an open.

10. Error in the *assembly*: You have a cold solder joint between a lead of the IC and its pad. The most common cause of a part not working as expected is a lifted lead or other bad solder joint.

11. Error in the *test setup*: The input voltage to the SMPS needed to be 7 V, but you applied 5 V from the external power supply.

12. Error in the *measurements*: The scope cable added a 100 pF capacitive load to the crystal and damped out its oscillation. Or you have a loose or poor connection and are not measuring the signal that is there.

13. Error in the *code*: Good luck.

Any problem in any category can be enough to keep your widget from working.

After the board is assembled to some level and something is not working as you expect, there are generally six obvious places to test to try to find the root cause:

1. *Check the power rails.* Start at the source of the power and measure each node. Verify the DC voltage expected AND the ripple noise level expected is present. When possible, measure the current draw from the power supply and compare it to what you expect. Try to isolate some of the circuits on the board using jumpers to isolate the problems.

2. *Check the signal paths* at the inputs and outputs of all the signal pins. Trace the signal path through each device. Verify the frequency, peak-to-peak values. Check for continuity of the signal path, from the TX to the RX and through the entire circuit. Isolate the circuit when possible.

3. *Check the voltage on each of the pins* of each component accessible. You are looking for hints at what might be a problem. Are all the voltages and signatures for the various pins what you expect to see?

4. *Check the noise on signal lines.* If there is a quiet low and high that can be used, you can measure the cross talk on these quiet lines directly. Assemble the power rail components and test the voltage.

5. *Monitor the power rail noise while the part operates.* Often, the voltage noise on the power rail can be a useful diagnostic for what might be going wrong in the circuit. Knowing the expected power budget will give you insight into what is "normal" and what might be an indication of a problem somewhere.

6. *Look at all the solder joints under a microscope for poor solder connections.* Sometimes it is obvious, as when the lead is lifted. Sometimes it is less obvious, as when the lead is on top of the pad, but the solder has not reflowed to the lead and there is an oxide layer between them. This is a cold solder joint. If any joint looks remotely suspicious, reflow the joint by first applying a lot of solder flux and then reheating the pads with a hot, clean soldering iron tip.

25.6 Troubleshoot Like a Detective

When the prototype fails a test, like the power light does not go on, or there is no signal out of an I/O pin, or an output does have the signature you expect, or the prototype is not seen on the USB port, it's time to move to the debug phase.

Watch a detective at work. When they encounter a dead body, their goal is to find who murdered the victim. They look for clues. They interview suspects. They create a murder board on which they post all the evidence and suspects and their plan to find the guilty suspect.

It is exactly the same process many different professionals use when dealing with their problems, such as:

- ✓ A doctor
- ✓ An auto repair person
- ✓ An appliance repair person

An engineer uses essentially the same process to find the root cause of a problem and fix it at the root cause.

This process is variously referred to as:

- ✓ Debugging
- ✓ Troubleshooting
- ✓ Forensic analysis
- ✓ Diagnosis

When you are assembling your board and the circuit is not working the way you expect, what do you do? How do you find the problem?

For example, the 555 timer should be outputting a 1 kHz square wave. But maybe it is outputting a 100 kHz, 95% duty cycle pulse pattern. What is wrong and how do you fix it?

Detectives look for clues, they find suspects, and then they figure out what questions to ask each suspect to either establish an alibi that says this suspect cannot be the killer or is consistent with the suspect being the killer.

You look for every clue to point to a suspect and then ask questions to check for an alibi.

Where do you look for suspects who might have killed the 1 kHz signal? Could it be the wrong timing resistors or capacitors? Could it be one of the leads of the 555 timer is not soldered to the pads? Here are two suspects. What questions would you ask them to help establish the guilty party, or exonerate one of the suspects to move on to another?

- ✓ You could inspect the leads under a microscope to see if there is an assembly defect.
- ✓ You could resolder the leads to correct any invisible assembly defects.
- ✓ You could measure the resistance of each timing resistor to verify its value.
- ✓ You could measure the capacitance of the timing capacitor to verify its value.

Each of these tests is suggested based on an assumption of a root cause.

When you think about something to look for, or something to try to fix the prototype, you are unconsciously making an assumption about what you think the root cause is and then addressing that potential root cause.

Bring these thoughts closer to the surface. Consciously articulate the specific possible root cause you are thinking and how you would test for this specific root cause. Walking through this process and formalizing it will help train your intuition into looking for a root cause and developing a test for that specific root cause.

Once you have verified the root cause, fixing it is straightforward. You always want to figure out a way of fixing the problem in the current prototype so you can move on to find and fix other problems. But sometimes, the fix can only be implemented in a redesign.

Once the guilty suspect is narrowed down, what could you do to fix the problem based on the root cause? Don't jump ahead and try lots of different fixes without first identifying and confirming the root cause.

If you just try a bunch of things and the problem goes away, you will never have confidence you fixed the real problem and you will not have learned anything in this process you could use for the future.

There are some general principles that apply to all of these diverse applications, and some specific special cases.

> *In general, the better you know how your product is supposed to behave (rule #9), the better you will be able to identify when it is not working, and what the reason is. There is no substitute for studying manuals, datasheets, and knowing the schematic intimately.*

25.7 Trick #1: Recreate the Problem

There is no substitute for reading the manuals or datasheets of all of your parts. The better you understand your system, the better you can identify the small hints when it is not working and can guess a possible root cause.

There are four tricks you can sometimes use to help you find the root cause.

Try to think of how you could re-create the problem on purpose. The problem is, you expect to see a 5 V signal coming out of a pin, but you see a 2.5 V signal. What is the root cause?

Ask yourself, if I wanted to turn a 5 V signal into a 2.5 V signal, how could I do that? How could you make this happen on purpose?

You could build a voltage divider with a few resistors. Maybe there is a voltage divider created accidently in your circuit? This could be by having your digital pin set as a pull-up and the 10k resistor on your board as a load.

Look for a 10k load on the output and check your code to see if the digital pin is set as an OUTPUT or as a PULLUP output.

When you see a specific behavior, think about how you could re create it if you had to, and then check the prototype for those features you would have added to create this behavior.

25.8 Trick #2: Seen This Problem Before?

Keep a journal, a list of the problems you have found: the symptoms, indicators, hints for the problem and its root cause. Each time you encounter a problem that has happened before, add it to this list.

You should have a list of symptoms and causes. If you see oscillations on the output of an LDO, this usually means you forgot the 22 uF decoupling capacitor, or the capacitor is not large enough. Maybe you thought it was 22 uF, but it is really 22 pF.

When possible, take a functioning board and cause known problems, like remove the filter or decoupling capacitors, replace a resistor value with a 0 ohm jumper, or pull one of the signal pins off the pad. Observe the signature of the failure on the board and make note of the signature and its root cause. Add this example to your list.

All measurements should be stable to mechanical agitation. If you notice that the signal displayed on the scope changes as you move a wire, or jiggle a connection, or flex the board, or tap a connection, this is usually an indication of either a loose wire, a bad solder joint, or a broken trace on the board.

A common problem with a 555 timer is to short the pin 5, which is the control voltage pin. Normally there is a 1 uF capacitor on the pin. If you remove the capacitor, keep the pin as a no connect.

If used as a monostable, the input trigger pin needs to be a voltage, pulling pin 2 low for a time short compared with the expected output pulse. If this is not the case, the output pulse will not be as you designed it.

It is especially useful to make perturbations in a simulation tool. This way you can observe a specific signature to the signals when you create a known problem such as poor termination or excessive cross talk to an adjacent trace.

25.9 Trick #3: Round Up the Usual Suspects

This is a line taken from the famous movie, *Casablanca*. If you have not watched it, add it to your list. Watch this scene here: https://www.youtube.com/watch?v=HXuBnz6vtuI

There are probably ten common problems or usual suspects, which are likely the root cause of most problems, such as:

1. A loose wire, a cold solder joint, an unsoldered component, or a broken trace on the board. Look under a microscope, add solder flux, and reflow the solder joints. Fix the bad connection.

2. Error in the circuit connectivity — TX to TX or D+ to the incorrect input.

3. Wrong capacitor or resistor value — remove the part and check with a meter.

4. ESD damage to the part — replace the part.

5. The part was thermally damaged during soldering — replace the part.

6. Not plugged in — possibility a bad connection or power supply not on.

7. LED assembled in the wrong polarity — test the LED with an LED tester.

8. Design error in the layout. Trace the signal paths to verify they connect where they are supposed to and refer back to the datasheet, manual, or a reference design.

9. A component is missing from the board. Check for any empty pads.

10. A jumper flag is not connected correctly.

Each time you encounter a problem and find the correct root cause, add it to your list. The more boards you debug, the more usual suspects you will have on your list to check.

25.10 Trick #4: Three Possible Explanations

One of my college professors, the late physicist Prof. John G. King, was famous for saying,

"When something happens, there is always a reason why. Any physicist can come up with one explanation for why it happened that way, but a really good physicist should be able to come up with at least three possible explanations."

When you see a behavior you did not expect, as practice, you should push yourself to come up with three possible explanations.

Each one becomes a suspect, which requires further interrogation. This means conducting a consistency test to look for clues or to eliminate a suspect or further interrogate.

This process of thinking of three possible explanation is just an exercise to push you to use your creativity.

Of course, the more you understand about your system, the more you have explored failure modes and thought about their signatures when they fail, the easier it will be to think of three possibilities.

The more problems you encounter, as long as you are able to pin down the root cause of the problem and confirm the root cause, the more likely you will guess the root cause for the next problem.

25.11 A Methodology

If there are no obvious clues to investigate, there is a simple methodology to follow to search for clues to help identify a potential root cause.

When you are troubleshooting, you are constantly thinking of possibilities. To help stimulate your thinking, it is useful to write things down. The process of documenting observations and what you did and your prototype's response will help stimulate some ideas.

It doesn't matter if you use a paper notebook (old school) or an electronic notebook like notepad++ or Microsoft OneNote or even a simple Word document. It's the process of articulating your thoughts that will get your intuition focused and maybe channel some insight.

This is why working with a teammate is so valuable. The process of articulating what you did or are doing will help stimulate your thinking to maybe realize some subtle connection that will give you insight into a possible root cause. As you explain to someone else what you are doing, insight and understanding often follows.

First, do a visual inspection. Look for clues. Maybe something obvious will stand out. Is there a break in a surface trace? Is there a bad solder joint with a lead lifted off the board? Is there a pad missing a component?

Always inspect the leads under a microscope to verify you have good solder joints on ALL the leads. A good stereo microscope will show a bad solder joint with the solder not wetting the lead, a lifted lead, or a partially reflowed solder joint.

Check the schematic against the datasheet. Did you connect the pins as they are supposed to be connected?

Start with the power rail. Measure the voltage on the power rail at the device pin. Use a scope so you can see what the noise might be. Look for the DC level and the transients that might be on a short time scale. Use a slow time base on the scope and gradually go up to as fast a time base as your scope allows.

If you do not see the correct voltage, like 5 V or 3.3. V on the power rail at the device pin, then gradually move back toward the source until you find the correct voltage signature. The root cause is a problem along the way in the power path.

If there is no power and the device is plugged in, and the power source when unplugged works, there may be a short between the power and ground on your board. This is why it is always a good idea to measure shorts and opens in your bare board before components are assembled.

Or, there could be a bad component that is shorting out the board.

If the pin connections look good and the power is good, check the signal paths.

Look at all the input signals with a scope. Do they have the signature you expect? What is the noise level?

Trace the signals through from their inputs to their outputs.

If you have isolation switches, turn off parts of the system. Take advantage of the switches, indicator LEDs, and test points you integrated into your prototype.

25.12 Forensic Analysis

The fastest and most efficient way of fixing the problem is to first identify the problem and then find the root cause.

Here is the checklist of actions to consider taking at bring-up:

1. Before assembly:

    a. Do a visual inspect of the board to look for obvious defects like broken traces, shorted traces, or missing pads.

    b. Measure the resistance between different power rails and ground with a simple 2-wire DMM to verify their isolation is larger than 1 Megohm.

2. After each component is assembled:

   a. Do a visual inspect to look for obvious bad solder joints.

   b. When some joints look a little suspicious, like a dull gray instead of shiny, apply solder flux to the leads and reflow the solder so that it clearly wets the leads and the pads.

   c. Any solder shorts can be cleaned up by reflow and letting the surface tension pull the excess solder to either lead. Alternatively, copper braid or solder wick can be placed over the leads, with a liberal amount of solder flux added, and heated with a hotter-than-usual soldering iron. The solder will be sucked into the copper braid and excess solder removed.

   d. Check for any missing solder joints or for pads that do not look like they have adequate solder, especially on a part with many leads.

   e. Generally, it is easier to assemble the smaller parts first, then the bigger ones.

   f. Be aware that a hot air gun may damage plastic connectors already placed on the board.

   g. Always double check the orientation of parts that are polarity sensitive.

   h. DO NOT solder any parts on the board while it is powered on.

   i. Silk screen markings can make it so much easier to double check you are adding the correct part — add the component name, the component value, and the component polarity often, in close proximity to the pads.

3. The power path:

   a. Check the current draw from the power supply. This is one reason to use isolation jumpers to turn off some parts of the circuit so that the power draw can be measured. This is done using an external power supply that allows measuring the current draw. Is it what you expect?

   b. Trace the DC voltage level at each stage of the circuit. Is the DC level being distributed where it needs to be?

   c. Look at the noise on a scope to verify the voltage level is what is expected at each stage and the noise is low.

   d. Verify the decoupling capacitors are positioned in close proximity with low loop inductance to the IC they are decoupling.

4. The signal paths:

   a. For each component, verify the signal you expect to see is what is coming out and it is connected between each output and into each input. Use the sharp tip of the 10x probe to probe the leads of packages or pads of components, or untented vias.

   b. For digital signals, select one as a trigger and look at the corresponding reaction in other pins.

   c. Consider modifying the microcode to change the signals on other pins.

   d. Measure the clock frequency to verify it is operating at the expected frequency.

   e. Trace the routing from an output to an input. If the routing is incorrect, pull a lead from the pad and connect it to the right place with AWG30 wire. This is called an engineering change wire, green wire, or white wire change. Check out an example of a white wire change on a rev 1.0 production board in **Figure 25.2**.

Figure 25.2 An example of a white wire change on a production board. Source: https://byuu.net/cartridges/boards.

5. Quiet high and quiet low noise:

    a. Write the microcode to make one line switch and use this as the trigger. This pin will trigger the scope so you can look for synchronous switching noise on the power rail and the ground pins.

    b. A quiet LOW pin is an I/O that is set by the microcode you write to be outputting a LOW. Normally, this voltage will be 0 V. Any voltage measured on this lead is cross talk from other signals switching.

    c. A quiet HIGH is an I/O that is set by the microcode you write to be outputting a HIGH. Normally, this voltage will be the Vcc rail voltage. This is a direct connection between the power rail on the die to the lead on the board. This is a direct measure of the power rail voltage noise on the die.

d. Monitor the quiet LOW and quiet HIGH to check the switching noise when other I/Os switch.

6. If you can verify a specific part is not reacting with the correct outputs or responses to inputs:

   a. Inspect its solder joints.

   b. Consider reflowing any suspect joints.

   c. Verify it is placed on the board correctly.

   d. Verify from the schematic that the pin connectivity is correct.

   e. Verify the correct power and signals are going into the device by measuring the voltage on different pins.

   f. If there is a crystal, measure the voltage on one of the crystal leads to verify there are oscillations at the frequency you expect. Use a 10x probe.

   g. If the device is connected correctly and assembled correctly, it could be a bad part. It could have been damaged by ESD or thermally in the assembly operation.

7. Replace a potential bad part only *after* you have exhausted other possible explanations. Maybe it was shipped as a bad part, maybe it suffered ESD damage or maybe it was thermally damaged.

   a. Using a hot air gun, slowly heat up the part and its pads and pull it off the board using tweezers when the solder joins have melted.

   b. Clean up the pads with copper solder wick and a soldering iron, using solder flux to help the wick suck up the solder.

c. After the pads are cleaned, add new solder paste, reflow the solder paste to let it flow over the pads, and replace solder to the pads.

d. Add solder flux to the pads and place the part on top of the pads.

e. Hold it in place while reflowing using the hot air gun.

f. Repeat the test for the correct power and signals.

25.13 Coding Issues

Debugging software can be notoriously difficult. Some problems are only hardware related. Some problems are only software related. But some problems are both. They involve the interactions of the code running the hardware, often referred to as firmware. This is the world of physical computing.

The first step is to try to separate the problem.

Is there a working hardware evaluation platform you can use to debug the software? Can you emulate your hardware system on a software platform?

Is there some simple code you can run that will test the hardware? The simpler the better.

Is there a golden system you can test your hardware on and look for specific signals?

Can you develop the right test vectors to test the hardware as it operates under some code?

The more modular you can write your code, the more easily you can test each module to isolate a potential problem. The easiest to debug software is written as function, written with testing each function separately in mind. The main loop should only call functions. If you can write your code modularized, it will be so much easier to debug.

25.14 Practice Questions

1. What does it mean for your prototype to work?

2. When do you establish the criterion for a working widget?

3. What are two features you might include in your prototype and not have in your production board?

4. What are three features you could add to your board specifically for bring-up?

5. If you board does not work, what are three possible root causes?

6. If the 3.3 V rail is showing 0 V, what are two possible suspects and what consistency tests could you do to verify or eliminate these possible problems?

7. An inverter is being used as an RC timer. The expected frequency is 1 MHz but you measure 1 kHz. What are three possible explanations to investigate?

8. What are three possible reasons a specific component on a board does not work?

9. If you exhaust all possible explanations for why a transistor does not work as expected, what might you try?

10. What does it mean to round up the usual suspects?

Chapter 26
Step 7: Documentation

The last step is documentation. This can range from writing a project report, to writing a product manual, or even writing a technical article about your design.

Most documentation, whether a final report, a publication, or even a user manual has two important purposes.

First, by articulating what you did or how the system behaves, you are forced to think through and reflect on the details. The process of documenting your work helps you think through the details and why things went the way they did. The more you think about each step of your progress, the more you can analyze the consequences of your actions and what you might consider doing differently next time.

The second purpose of creating a document is to transfer your vision to another person or group. You've designed a product or created something new. The documentation you create is your way of transferring what you have learned or how this product performs to someone else, so they can benefit from your insights.

Independent of any publication, you should be keeping a notebook with your personal observations and lessons learned. What errors did you encounter that you will not have to make the next time, and what solutions did you find?

Regardless of the purpose, every report is really a story. You've finished the project, you know the complete story arc, now you can put the pieces together and rewrite the story based on the message you want your readers to walk away with.

Do you want to highlight the problems you encountered and overcame?

Do you want to focus on the journey you took, from start to finish?

Do you want to show a path for success from start to finish so the next person following you can get to the end point more quickly?

Do you want to highlight the lessons you learned so that you will not forget what to do differently in your next design?

When you are working on this project, you should be thinking about pictures or measurement that would have been useful to illustrate your story.

Is it an assessment of what went right and what went wrong? The process of articulating these features helps in focusing your attention on what you did.

The more documentation you do, the better you will anticipate what you should record as you go, so as to have it available in final report.

Never include data or a scope trace without your analysis of the "so what?" What does it mean? This is the most important part of the documentation.

Chapter 27
Concluding Comments

All electronic prototypes are complicated systems. There are far more ways for things to go wrong than to go right. This is why you will increase your chance of success by designing more like Ralphie's mom and less like a Colorado Bro.

An important theme mentioned often throughout this book is that successful designers consider all the things that can go wrong, anticipate them and prevent them from occurring. If you are called a worry wart by your colleagues, wear the moniker with a badge of honor – you are doing it right.

The more you understand about the details of your components, your circuits, and the expected performance of your product, the quicker you can bring it up, finding and fixing potential problems along the way.

If you expect to design and build more than one electronic protype in your career, start early adopting good habits in the circuit design and layout that will give you the lowest noise. You may not need this low noise performance in your current design but absolutely guaranteed, a future product will benefit from your good design habits that result lower noise.

Keep in mind, the ultimate goal of any product design process is to deliver a product with the specified value to the end user at an acceptable cost, on time. Take every opportunity to reduce your risk and increase your chance of success in all of your future prototype projects.

Chapter 28
About Eric Bogatin

Eric Bogatin is currently a professor at the University of Colorado, Boulder in the ECEE department, technical editor of the Signal Integrity Journal, and a Fellow with Teledyne LeCroy.

Eric received his B.S. in physics from MIT in 1976 and M.S. and Ph.D. in physics from the University of Arizona in Tucson in 1980. He has held senior engineering and management positions at Bell Labs, Raychem, Sun Microsystems, Ansoft, and Interconnect Devices. He has written nine technical books in the field of high-speed digital engineering and presented classes and lectures on signal integrity worldwide.

He can be reached at Eric.bogatin@colorado.edu. His faculty website can be found here: https://www.colorado.edu/faculty/bogatin/ .

CPSIA information can be obtained
at www.ICGtesting.com
Printed in the USA
JSHW062124170922
30648JS00001B/1